The Great Moon L

Or … Dealing wit

CW01429141

Peter Bassett F.R.A.S.

Astronomy Roadshow Publishing
Further paperback copies and an e-book version with direct
internet links can be purchased from…

www.astronomyroadshow.com

Supporting Website; www.moonlandinghoax.org

There are many Internet links of video clips etc related to
many of the chapters. Just search the moon landing hoax
website for the relevant page.

**This is an economical Black & White version. Full colour is
also available from either of the above websites**

Every major space mission, manned or unmanned has a patch to represent it. Manned mission patches are normally designed by the crew. This tradition continues today.

Contents   Page

This is a pure scientific explanation of where the conspiracy supporters are going wrong. It is important to know the difference between real science and guess work. Nothing new will ever be invented or discovered otherwise. It is okay to have your head in the stars as long as your feet remain on the ground. This is the only way real progress can be made.

The supporting website is
www.moonlandinghoax.org

Clips are shown on the website where a book cannot. The entire documentary that commenced this hoax business by Fox TV is included.

Any text marked in BOLD RED (BOLD for the B&W edition) is intended to show what a typical hoaxer says about the point in question.

Chapter 1   Introduction

Many people around the globe must now have heard of the idea that the six Apollo moon landings carried out between 1969 and 1972 were a hoax. A massive conspiracy was created to dupe the entire world that American astronauts were walking around on the moon. But instead they were in the safety of a studio just as in the movie Capricorn One; a great film of a Mars Landing Hoax with guest star Telly Savalas – Kojak (the series ran from 1973-1978).

This book is intended to show the clear difference between real science & investigations over inventive imaginations. It will not be a comprehensive guide to the history of Apollo as this has been covered in extreme detail by other publications.

Readers may use this book as ammunition against the conspiracy supporters. Some of the arguments forwarded by them can be confusing at first and difficult to argue against, the counter argument is displayed in this book with the homework completed.

With the age of the Internet, it is so easy to publish absolutely anything and get a message across to millions of people. For those that are not experts in any particular field, it is relatively easy to convince them of any particular belief.

The aim of this book is to define the difference between reality and fiction. This investigative skill can be used in many walks of life and connects with various occupations;

Police
Teaching
Journalism
Armed Forces & Politics
Inventors and general science research
Analysing a business opportunity

Some companies use group discussions on a mystery of some form to help choose candidates for a specialised job. This book may help to train your mind to achieve such tasks with ease.

NASA even once developed a hypothetical scene of being stranded on the moon and you had to choose items to take along on a dangerous trip to a moon base. I was able to correct several flaws in their own guidelines by using the same analytical methods.

Science is a complex subject and scientists do make mistakes as the Universe is so strange. But false conspiracies and bad science like this example make their job harder still.

"These people are doing a terrible disservice to the youngsters in education today" ... Buzz Aldrin; Apollo 11.

"Americans never landed on the moon? You might as well say America was never discovered!" Prof Brian Cox.

(Quotes from other Astronauts / Cosmonauts / Taikonauts / Scientists are very welcome for future editions).

This book is dedicated to the bravery and skills of all the Astronauts, Technicians & Scientists involved with project Apollo and to a very special friend for 20 years...

Dennis Culver of Huntsville, Alabama, USA. He passed away in Aug 2012. He loved Huntsville, home to Wernher Von Braun (father of modern rocketry), and was a dedicated member of the Von Braun Astronomical Society. We spent many hours discussing this moon hoax subject and both agreed that something like this book / website should be developed to put the bravery of the astronauts back where it belongs.

Most of us lead very busy lives so this book has a structured order to use as a quick reference guide as well as read cover to cover. No section contains dozens of acronyms that require memorising from a previous chapter.

Chapter 2   In the beginning

Having watched the Apollo missions unfold one by one in the 60's & 70's and seeing Mum squeezing the laundry dry using a wooden mangle in the garden, the two forms of technology seemed complete opposites. It was hard to believe that we really did have men walking and then driving around the moon while most of us tried to keep warm by a log or coal fire and TV sets often burst into flames. Landing on the moon just seemed too far advanced for the times.

Today we have the same situation but on a smaller scale now that modern technologies are more widespread, reliable and affordable. However there are still sections of the human race living in mud huts, igloos and tepees as astronauts conduct experiments passing overhead in the International Space Station some 300 km up travelling at 5 miles per second.

So it is understandable to a certain degree that some people just can't accept that some of these amazing achievements occurred at all. I will allow this point that commences doubts in some minds as to the legitimacy of Project Apollo. But as this is investigated fully a section at a time, then the truth is revealed and hopefully such thoughts pushed aside to allow individuals to get back on track to real education. It is this book's mission to untangle minds and set a more solid course in solving the world's real problems.

Chapter 3 The Documentaries

Several documentaries have been produced directly relating to this moon landing conspiracy. Bill Kaysing was always at the heart of this rumour. He did work for Rocketdyne as he claimed but Rocketdyne didn't even produce the Lunar Module engine he mentions; it was Space Technology Laboratories and later bought out by Grumman. He didn't work as a rocket engineer, he gives the impression he was; Bill was a filing clerk and writer of technical papers from the engineers; his next career was looking after stray cats. He even left NASA five years before Neil stepped on the moon. This is one example of how to get to the truth behind Bill's constant ramblings. Other characters include Ralph Rene, Marcus Allen, James Collier and many others.

To give Bill some credit for his suspicions, he did witness and re-type up many reports about experiments with failed engines; especially the F1 that was used on the Saturn V. It probably gave him the impression that the technical requirements just weren't being met. Several hundred modifications later, the engine passed man-rated tests and finally did make it to the launch pad. By then Bill had left NASA.

When I first saw 'Did we land on the Moon?' by Fox Television, I was screaming at the TV set 'How dumb are you guys?' and 'You are embarrassing yourselves at publishing this nonsense'… I don't think they heard me.

This book and website is the result of that screaming. Fox TV continue to embarrass themselves by screening such poorly researched material. Perhaps they will realise their error someday and withdraw it. Can you then guess as to what the conspiracy supporters will say next? "See! They do have something to hide and the US Government banned it from further transmissions or they will murder the boss of Fox TV."

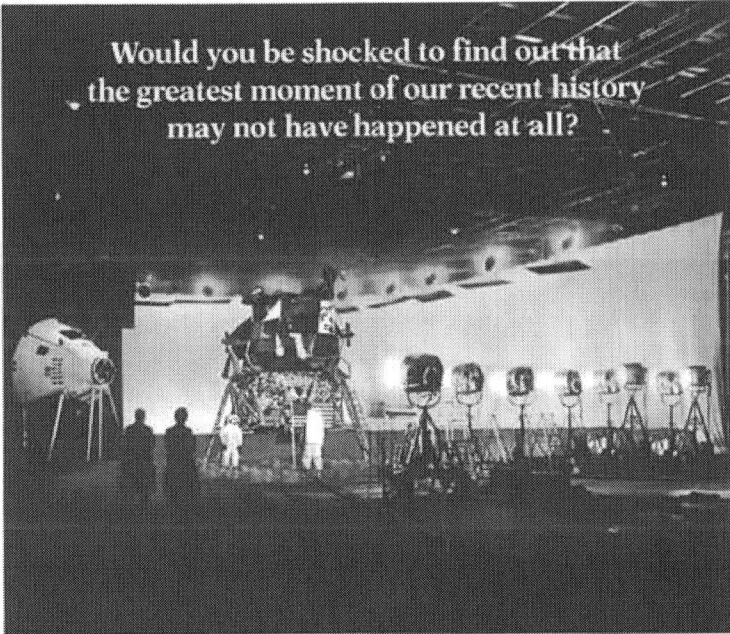

The movie poster for Capricorn One.

Chapter 4   The Books

Several books are available on the Moon Landing hoax; 'Dark Moon - Apollo and the Whistle-Blowers.' published by Aulis Publishers is probably the most well known. It connects the moon-hoax with pyramids on Mars, Wigwams, Stonehenge, Atlantis, the Sphinx, Neanderthals and Crop Circles all in one 'neat' theory. If you want a good laugh then please do get a copy. Amazon is where I got my *used* copy, not a penny went to the authors. I would hate to feel these people are gaining even more money through this advert. David Percy & Mary Bennett wrote this as self publishers.

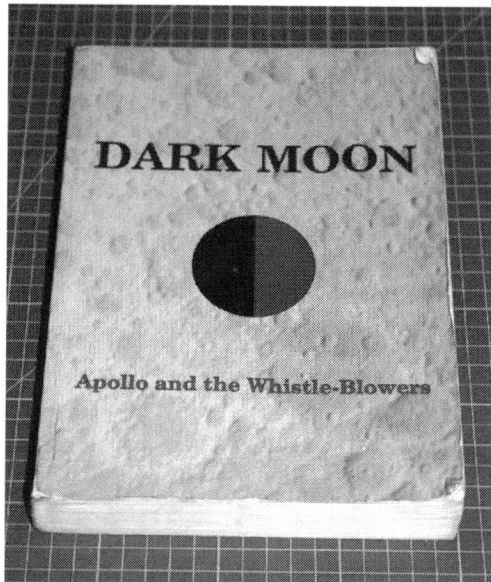

On just two particular pages, the author connects Aliens, a baby's hand, snowflakes, and Radio Astronomy... David Percy is a genius. It's the same throughout the whole 500 page book. On page 26, he shows a picture of two astronauts with slightly different length shadows on the moon. He claims it's due to a close studio spotlight. Has he thought it could be that the astronauts were different heights or more likely, standing on ground that wasn't level? This is the limit of his intelligence. As

10

Dr Patrick Moore often said "Some people may take this with a pinch of salt, I would take it with an entire salt mine!"

The very first book about the subject, We Never Went to the Moon: America's Thirty Billion Dollar Swindle, was written in 1974, two years after the Apollo Moon flights had ended. In 1976, by Bill Kaysing (1922–2005), a technical writer hired in 1956 by Rocketdyne, the company which built the F-1 engines used on the Saturn V rocket, despite having no knowledge of rockets or technical writing. He served in the technical publications unit at the company's Propulsion Field Laboratory until 1963. Kaysing's own book made many allegations, and effectively began discussion of the Moon landings being faked. His book claimed that the chance of a successful manned landing on the Moon was calculated to be 0.0017%, and that despite close monitoring by the USSR, it would have been easier for NASA to fake the Moon landings than to really go there. The reverse is actually true.

The Flat Earth Society was one of the first organizations to take up this idea and accuse NASA of faking the landings, arguing that they were staged by Hollywood with Disney sponsorship, based on a script by Arthur C. Clarke and directed by Stanley Kubrick. Folklorist Linda Degh once suggested that writer-director Peter Hyams 1978 film Capricorn One, which shows a hoaxed journey to Mars in an Apollo craft, may have given a boost to the hoax theory's popularity. She correctly noted that this happened during the post-Watergate era, when American citizens were in a mood to distrust government claims. Dégh writes: "The mass media catapult these half-truths into a kind of twilight zone where people can make their guesses sound as truths."

The Moon Landing Hoax by Dr Steven Thomas; Available as a Kindle book (Amazon) or in paperback. This author is a geologist but seems to be an expert in photography and rocket motors instead. Less than 100 pages, it is written with 'wishy washy' arguments for a scientist. The simple answers to all 'problems' with photography etc are all found in this book. Again I purchased a *used* paperback copy.

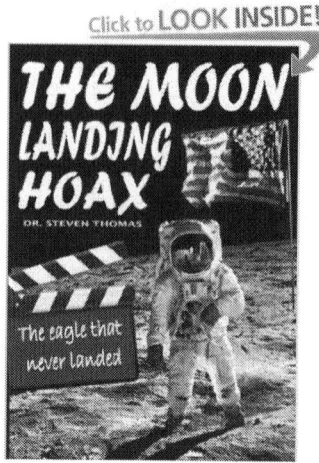

I won't list all the books available that support the hoax, all anyone needs to do is search 'Moon Landing Hoax' on Amazon or similar book sites and all are revealed. Such author's are totally convinced they are correct about their claims and you may feel that strongly wish to contact them. I have tried many times, but never gained a single reply for years.

Contact details for this book are on our website www.moonlandinghoax.org . Replies will be made if they are polite and straight to the point; our time on this earth is limited but my mission is to get science education back on course instead of 'grey area nonsense.'

Chapter 5   Order of Investigation

To investigate a crime, business failure and yes even a conspiracy, there needs to be a set order in which to attain the truth. The fastest method is always to analyse the simplest 'facts' first. This in turn already will give a strong indication of the final result. The moderate details come next and then the fine picky loose ends that require a tidying up for the final conclusion. An open verdict can exist if some of the evidence alters over time or new evidence comes to light.

The website associated with this book will display new material as the 'evidence' of a hoax materialises. Any reader may submit information from either side of the argument at any time. This is the future of book publishing.

Having watched the popular TV detective series 'Columbo' from the 1970's, I was fascinated as to how each carefully planned murder was solved. Peter Falk, playing Columbo noted not the overall impression of the murder scene as other detectives did, but the little details instead. This is how I structure conspiracy investigations. Hoaxers prefer to bash on one point after another without taking a breath in between. They need to slow down and study each point with an expert in that field to ensure they have the right answer; they never do.

The structure of this investigation is based upon a set procedure…

Every single mystery or complex task of any type should be broken down into small segments.

List them in a rough order from simplicity toward complexity.

Investigate each point by using several sources of real experts. Be honest and realistic about the answers even if you originally disagreed with it.

Chapter 6   Flapping Flags

We are often told by the hoax supporters that flags cannot flap around on the moon because there is no wind! Sometimes a moon flag is seen flapping around when no astronaut is anywhere close to it. How can this be if there is no wind either? The astronauts must have been on earth and a breeze from somewhere made it flap around.

As the flag is in a vacuum (no air) on the moon; there is also no air resistance either. Shake a flag in a hall and then leave it, it will stop within 3 or 4 seconds due to the 1000mb air pressure surrounding it. The energy you put into the flag is lost to the air. If a flag is shaken on the moon, then air pressure doesn't exist to slow it down. It will continue to flap around for ages. So can flags flap on the moon? Yes!

Drop two flat pieces of paper of the same size. They both flap around until they reach the ground. Repeat the experiment after screwing one of the sheets up into a tight ball. Now the screwed up piece will hit the ground first and fall in a straight line without flapping around. This demonstrates the power of air resistance. On the moon, both sheets of paper will reach the ground at the same time regardless of whether it's screwed up or not.

If an astronaut pulls on the washing line and let it go, it would swing for hours or even days due to momentum. There will be no air resistance to slow it down, just a little friction on the poles where the rope is tied.

Illustrations by Brian van de Peer of Strood, Kent, UK.

The astronauts have gone home, but the washing line is still swaying due to a lack of force to stop it.

This is probably the simplest demonstration of how the hoax supporters imagine the properties of a vacuum rather than examine. Any physicist will confirm this simple point.

Chapter 7   Flags lit from both sides?

Some images on the moon show the US flag illuminated from one side and yet shadows are on the same side. So where is the extra light coming from? A spot light in the studio?

Apollo 15 image with Dave Scott saluting the flag.
NASA image obviously.

This and many other images have the sun behind the flag from the camera's perspective. The shadows are thrown forward, everything is backlit. So how does the flag appear to be lit from the front too? Ahah, a spotlight is concentrated on the flag to enhance its status; it is the US flag after all.

Is the flag made of heavy duty opaque cloth? No! What were they made of and how come they contain creases too?

About three months prior to the July 1969 Apollo 11 mission, Jack Kinzler, head of technical services at the Johnson Space Center, Houston was to design the flag assembly. Kinzler came up with the idea of inserting a horizontal pole through a hemmed pocket in the top of the flag to support it and make it appear to fly on the airless Moon as it would in the wind on Earth, from the memory of his mother hanging curtains. He also suggested, designed, and oversaw the creation of the commemorative plaques fixed to the Lunar Modules.

The flags themselves were simple, standard government supply 3-by-5-foot (0.91 by 1.52 m) nylon flags were altered only by sewing the top hem, its packaging, tolerance of environmental conditions, and means of deployment presented minor engineering challenges. The horizontal and vertical poles were each made of one-inch aluminium tubes in two telescoping parts. The total height of the flagpole was limited by the astronauts' 28-inch (71 cm) minimum and 66-inch (170 cm) maximum working height reach limits in their spacesuits inflated in vacuum. The flag cost $5.50, and the tubing cost $75. Coloured wires were sewn across the flags so when unfurled, they would take on a crinkly effect rather than a boring flat appearance as my picture below shows.

Well here is the simple and correct answer; the flag was made of thin nylon and was translucent. Strong light passes through the material. The larger flag is exactly the same size as the flag purchased for the moon. Anyone can recreate the same experiment.

Lesson learned; Science is about being able to repeat experiments the world over and beyond to get the same result rather than guessing.

Chapter 8   No Stars! Where are they all?

Our skies are blue due to the scattering of sunlight by Nitrogen & Oxygen. The sky on Mars is pink due to dust blown around in a Carbon Dioxide atmosphere. The moon's sky is dark day and night as there is no air at all. This means stars should be in every picture taken on the moon. The astronauts should also be able to see them; some said 'No' some said 'Yes.' How can this be?

The stars can shine visibly in a vacuum even during the day. To see and photograph them is another matter. One can stand in the middle of London on a clear night and see no stars at all, try to photograph them with any camera handheld that's not set for night photography and it's very unlikely going to capture a single star even though they are there.

We will investigate the 'seeing stars' part first. Stand under a streetlight, or in a floodlit parking lot, near a 1kw security light. How many stars can you honestly count?

The astronauts had a similar problem, they had landed every time during the lunar day. What is in the sky? The Sun. To protect their eyes from the bright glare, a gold plated cover was added to the visor on the helmet. Now reproduce the same experiment as before wearing gold tinted sunglasses / shades. Recount the stars; be honest please.

Well that answer was simple enough. Let's turn to the photography. During daylight on earth or in space, the brightness of the sun is constant as long as your distance from the sun roughly remains the same. With an Apollo moon mission, they travelled from Earth to the Moon; 150 million km to 150.4 million km from the sun, this is a negligible difference. The exposure of each picture will follow the same rules as if on earth.

When this picture of Chicago was taken, the stars were visible. But with a two second exposure, the skyline came out fine but where are the stars? A conspiracy supporter by his / her own argument must say that Chicago is fake! The correct answer is that the exposure should have been thirty seconds instead but the skyline would be ruined through over-exposure plus the street lighting would have turned the sky orange and drowned out the stars anyway.

In earth orbit the sky is dark there too. Not because there is no light, but there is no gas for the sun to illuminate; just the earth and the spacecraft itself. Therefore images taken in earth orbit always show a dark sky. But the exposure for the daylight section of each orbit will be the same as if in daylight on earth or the moon; around $1/250^{th}$ of a second. It does depend upon the sensitivity of the film used. The Apollo flights used film of 80 ISO. The same $1/250^{th}$ exposure was used for every shot taken of the earth and lunar landscape etc. For inside the craft, the speed would have been set to $1/60^{th}$ or so as the interior lighting would have been much dimmer compared to sunlight. There was no need to guess these figures or need exposure meters as these are constants and never vary.

Bruce McCandless, the world's first human satellite in orbit above the daylight side of earth. There are no stars in the image as the shutter speed was $1/250^{th}$ of a second instead of 30 seconds or more; 7500 times longer... but they were there!

Images such as these can only be taken with specific camera settings. A high ISO of 400 or above, heavy duty tripod, cable release to eliminate blurring and a long exposure of at least 30 seconds. But the astronauts used 80 ISO film, had no tripod or cable release, used 1/250 sec exposure as they were in daylight and didn't have the time for such images. $24 billion were invested to study the moon; not take pretty pictures of stars that can be done on earth. Both pictures by the author.

However there was one single exception. During the Apollo 16 mission, a 3" telescope was taken and deployed on the moon. It was designed to image stars in Ultra-violet. 178 Pictures were taken, and stars were recorded as a test for a larger telescope to be built on the Skylab space station. This one below shows the earth and stars in ultra-violet after a 40 second long exposure. The earth is overexposed as expected.

Chapter 9   Fake Moon Rocks

Hoax supporters claim that all the moon rocks were fake.
"Some are just meteorites collected from Antarctica" others say
"They were created in a Nuclear Laboratory." or even "Did you
know only NASA people are allowed to study the moon rocks?"
This is about all they say on the matter.

Every meteorite has a 'fusion crust' around it that developed on
its fiery journey through the Earth's atmosphere. None of the
moon rock samples have a fusion crust. They therefore cannot
be meteorites. If they were then why can't anybody produce a
meteorite rock today that is chemically & structurally identical
to a Apollo Moon Rock sample?

Most of the mass of every meteorite is lost on atmospheric entry
and wipes ALL the exterior helium-3 gained from solar
wind. Helium 3 and other exotic chemical traces that would
never be present on the surface of a meteorite but do exist in
Apollo Lunar Samples. The samples have been examined by
thousands of geologists and chemists around the globe who do
not work for NASA. Not one has cried "Fake" ever!

Every meteorite has a fusion crust as this one has. This is caused by entry into the Earth's atmosphere as the temperature reached thousands of degrees.

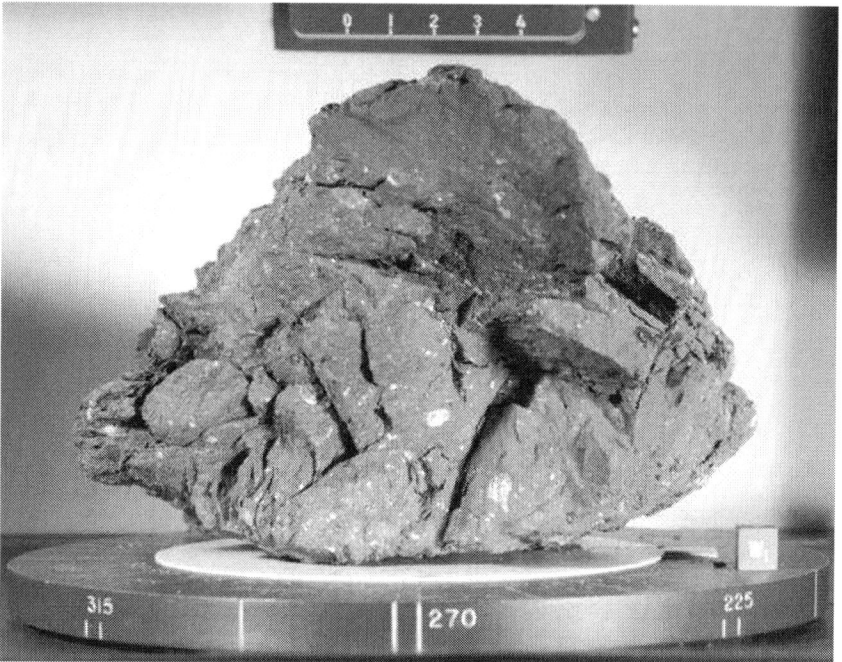

This material from Apollo 17 is Lunar Breccia. No fusion crust here or on any other sample. Plus Helium 3 is found plentiful throughout such samples but don't exist on the outer layers of meteorites as contamination from Earth will have eroded such properties. Ask such questions to a meteorite expert. If none are known to you, pop along to an astronomy club or even search adverts for meteorites on Ebay and ask the sellers.

Chapter 10 Wooden Rocks Now?

There was a case when a moon rock was given to former Holland Prime Minister Willem Drees during a goodwill tour by the three Apollo 11 astronauts shortly after their moon mission in 1969. When Mr Drees died, a rock went on display at the Amsterdam museum. At one point it was insured for around $500,000 (£308,000), but tests have proved it was not the genuine article; it turned out to be a piece of Petrified Wood.

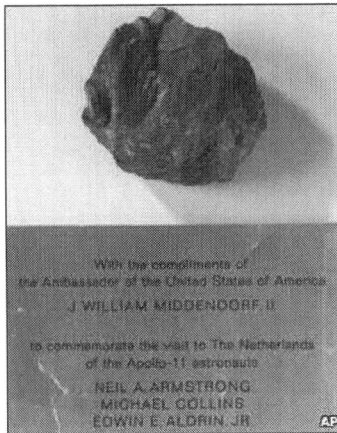

The Rijksmuseum, which is perhaps better known for paintings by artists such as Rembrandt, says it will keep the piece as a curiosity. "It's a good story", Xandra van Gelder, who oversaw the investigation that proved the piece was a fake, was quoted as saying by the Associated Press news agency.

"We can laugh about it."

The "rock" had originally been vetted through a phone call to NASA. The US agency gave moon rocks to more than 100 countries following lunar missions in the 1970s. A genuine moon rock was given to Holland and kept in Mr Drees's own personal collection. When Mr Drees died in 1988 after loosing his sight, officials handling his estate came across a rock and a card from NASA with the Apollo 11 astronauts names on it.

These legal officials were not geologists and simply assumed the sample was the moon rock in question and was put on display. No geologists were employed at the Art Museum. Four years passed when a geologist visitor recognised the rock as Petrified wood.

The Rijksmuseum in Amsterdam.

All the original moon rocks given out to be displayed were encased in Perspex but the petrified wood was not. This should have been a give-away but this was not known to the people involved in displaying the said item.

The museum has now a real piece of moon rock on display from the Apollo 17 mission. The original moon rock given to Mr Drees was never found.

The British National Space Centre hires out moon samples as well as Meteorites and Crystals. If one doesn't read the instructions, you may well be examining a meteorite thinking it was lunar material.

What about the other 400kg of samples that is examined by geologists & chemists around the world? Can they not tell the difference between petrified wood & moon rock? A fun story; but human error alone is to blame for the mix up.

Examine the full story rather than one sentence.

Chapter 11 The 'C' Rock

One particular rock somehow has a letter 'C' printed on
it from the Apollo 16 mission. This was marked to show
where an astronaut should be standing as actors have on a
stage when performing.

Yes this one picture out of thousands does seem to show a letter
C. But this particular image is a third generation copy after the
original negative has been through a photographic enlarger -
prone to dust and hair infection if not careful. Such items will
show up as a silhouette.

Note the quality of the image compared to the original – next
image. The contrast always increases on copies produced on
film. So the 'C' is just a fine hair laid on the copied image from
a dusty darkroom; nothing more.

Most of the published Apollo images seen in books etc are not from the original film taken on the moon; they are priceless and are handled to an absolute minimum. Images you see in print are often 2nd, 3rd or even 4th generation copies.

The entire Apollo film archive has now been digitised by high quality scanners to remove this problem of contamination by dust & hairs and decreased contrast quality. Such facilities were not available when this 'C' Rock picture was published.

Lesson learned; find the original source of evidence and not third hand as contaminants creep in and give false impressions.

Chapter 12   Project LOLA

**Top secret photographs have leaked out showing a modelled
moon landscape that was used for the Apollo orbital images.
These have never been seen before and released by hoax
whistle blowers proving beyond doubt all the moon pictures
were indeed faked!**

This artist is literally painting a copy of the lunar surface onto a
massive wooden sphere. The rendition of the moon was used to
train astronauts in recognising the lunar features for navigation
and landing.

So was the project a secret as the conspiracy supporters
suggest? Have these images just been leaked out?

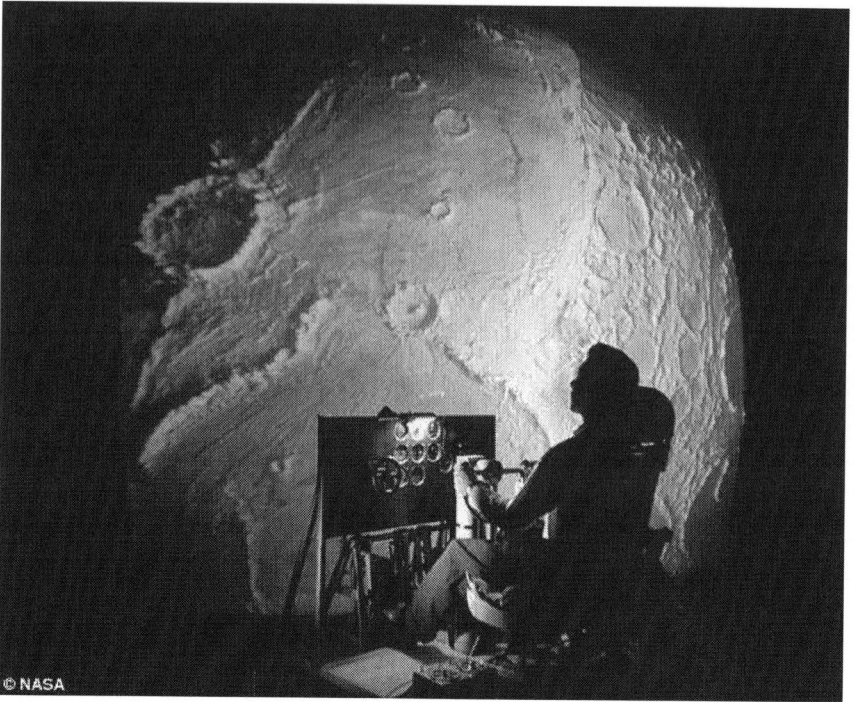

Introducing Project Lunar Orbiter & Landing Approach
Simulator (LOLA); a $2 million facility built at Langley
Research Center, Virginia. The lunar terrain detail was gained
by the American Lunar Orbiter series of probes that mapped the
entire surface in greater detail than ever before.

The central sphere contained Project LOLA.

If this project was such a secret, then can anyone explain why the project was published in a book called 'The Encyclopaedia of Space' in January 1969 and forwarded by a NASA astronaut - Scott Carpenter?

It was not a secret; the astronauts trained here for weeks at a time and has been published in many other places such as Life Magazine. Once again, question heavily these hoax people, they guess, assume & blunder their way around established spaceflight history that is already well documented. They just don't read enough books on the subject to discover such material and only take note of other conspiracy supporters.

Lesson learned; examine the true source of information and not assume any claim. Facts may be claimed to have been secret, but in reality it's just not widely read about.

Chapter 13 Those Wonky Shadows

Many shadows on the lunar photographs seem to be a little odd. The light from the sun is so distant that all shadows should be parallel on earth and the moon. But many photographs taken on the moon clearly show shadows in various directions. This would be correct if spotlights were used instead. Aha! The astronauts were in a studio and not the moon.

Hundreds of pictures taken on the moon show shadows acting very strangely. I won't show them all; this book will become too expensive and boring.

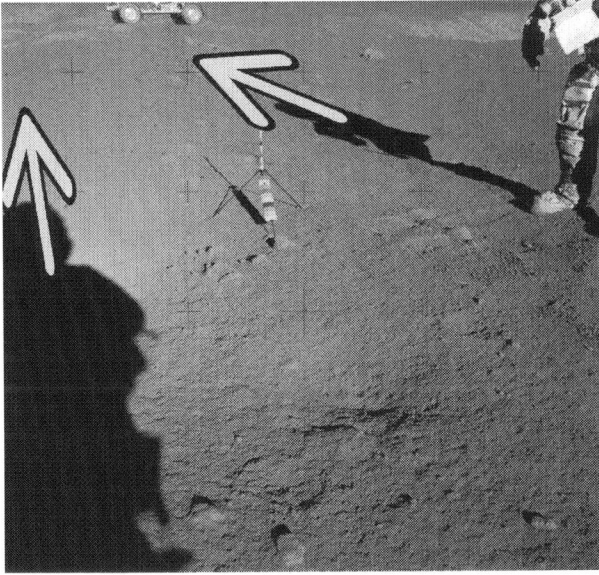

These shadows ought to be parallel to each other but they are not. The hoaxers claim that two separate spotlights, one on each astronaut, are angled inward by mistake in the above case. We seem to have indisputable evidence of foul play. Firstly if two spotlights were used there should be two shadows. I count one; please feel free to do a recount.

Here is an earth based answer; the shadows of my wife & I in Washington DC. As these shadows are heading off into the distance, they both appear to be converging but not actually converging. Anybody can observe the same effect and become obvious once you are aware of it.

I have set a perfect example in a more controlled way. Fence posts are of course upright and parallel – depending upon the skills of the fencer. The shadows they cause should produce parallel shadows; they do. But observe the following images…

Are spotlights required for such effects? Nope! It's called Perspective! Quite normal.

There are many examples of parallel paths, railway tracks, roads etc that show apparent converging or diverging lines.

Pictures taken from a low angle will always show shadows or rail-tracks etc diverging or converging. The ground can also be uneven or sloping; adding to the complication.

Crepuscular Rays are also parallel as the light is from the sun.

Lesson learned; Don't accept a complex interpretation when a simple answer is all around you.

Chapter 14   Sun Hotspots

**Several photos from each mission show a 'Hotspot' of light
on the ground. Aha! These must be caused by extra
concentration of light from a studio floodlight. All lighting
from the sun must be even. More evidence of a Hoax!**

This is Neil Armstrong on the moon with uneven lighting. The
picture below has the contrast enhanced by 30% to exaggerate
effect. Light from the 93 million mile distant sun will be evenly
spread so the lighting should be even too… Correct!

This time, in practice the result isn't quite so simple. The soil is
made up of tiny particles and they all reflect light. This effect is
simply a rough reflection of the sun. The same happens on
earth. We have all seen the effect but take no notice.

Different materials will reflect different wavelengths of light and at different angles. Look around on a beach or a flat road; the effect becomes obvious once you are aware of it. This is sometimes known as 'Forward Scattering' or 'Rayleigh Scattering.' The same effect happens on a molecular level not just sand, dust on the moon or ice. A green laser will reflect off dust particles so you can see the laser light more readily than most colours; a red laser won't work so easily. The daytime sky is blue due to the scattering of sunlight by Oxygen & Nitrogen.

This picture shows the same Rayleigh Scattering due to water. The moon's soil reacts in the same way. A full moon seen from earth is much brighter than just one day before or after due to an alignment with the soil particles and the Sun.

Picture by the author.

This 'spotlight' effect on the moon is quite normal. The same is observed on earth. Look out for it on roads, on a sandy beach, water or icy surfaces.

Chapter 15 Wires? What Wires?

When we study anything, we must always be honest with ourselves and don't imagine anything. Just observe what is there and never report what isn't observed. This should be applied not just to this subject but also to criminal evidence, ghost images, UFO 'evidence' etc. A well known TV series on ghosts have imaged 'spooky things' in mirrors & windows when it's just a reflection of the Infra-red Auto focus beam from the camera itself. The camera operator is told to pan past a mirror rapidly so when the freeze frame is used to 'study' the result, it's always a ghostly blur and won't show a clear reflected image of the camera Infra-red transmitter.

Such subjects can indeed be studied scientifically, but must not be contaminated with false imaginary evidence. This will lead any true advance astray and waste everybody's time, and even give the subject a bad name. Sticking to the truth will avoid this problem. It may sound dull and slow, but at least the truth is gained and not false hope.

The antenna on the backpack is an antenna; not a cable. Not a single image of a cable is ever witnessed. Yet the hoaxers claim over and over they are clearly seen.

The astronauts on the moon do leap around higher than on earth, and walk in a strange way we cannot duplicate accurately here on earth. But the hoax supporters, (David Percy in particular) claim this is due to 'wires' holding them up and lifting them on each step. Out of all the hours of footage of astronauts walking around on the moon, not a single wire is ever seen. Besides a single wire will allow the astronaut to rotate while off the ground, this never happens. A two-wire support would be needed as it does on stage acts. This would double the chances of the wires being seen, but is never recorded.

Anybody can download the finest quality scans of any Apollo image or video clip. Not one person has ever been able to enhance the image and reveal 'wires.'

Sometimes a flare of light is seen above an astronaut at the top of the frame. This is just a lens flare from the sun shining on the antenna mounted on the backpack, that's all! The hoax supporters continue to say they can see the whole wire. Not a single frame of 16mm film or photograph ever shows a wire. Some hoaxers insist the wire was removed from the film itself. Can anyone suggest how this was done to several million frames of film that was just 16mm wide with 1960's special effects? The wires simply don't exist!

Relevant clips are on the
www.moonlandinghoax.org **website.**

Lesson Learned; Observe what you can really see and don't imagine, no more.

Chapter 16   Stiff Spacesuits

The space suits / pressure suits are inflated for the entire body to breathe. Once outside, the zero external pressure would make the interior of the suit to puff up. The hoax supporters say that it would have been impossible to push buttons, operate equipment, bend down or do anything useful. They claim the internal pressure is too high and the suits will be too stiff.

If this is correct, can they explain how an astronaut today can go outside of the International Space Station (ISS), operate equipment, carry out repairs, bolt on new modules etc? The puffed up suits, according to their argument would make it impossible to build the ISS. Do they claim that all space-walks are fake not just on the moon? Then what about other projects such as servicing the Hubble Space Telescope, or earlier still the Skylab Space Station, or any of the Russian Space Stations. Do the hoaxers claim all these were fake because no astronaut can function properly outside in a vacuum? How far are they prepared to go with this?

Over 200 individual space walks (as astronaut James Newman above is demonstrating) were required to build the ISS. Some hoaxers are now actually saying the space station is a hoax too, the Hubble Space Telescope – fake! These people just don't know where to stop just because they never bother to study the workings of a space suit.

According to some hoax supporters all the missions that include space-walks by Russia, Britain, USA, Canada, and now China... are all fake because they claim spacesuits will not allow an astronaut to work outside in a vacuum.

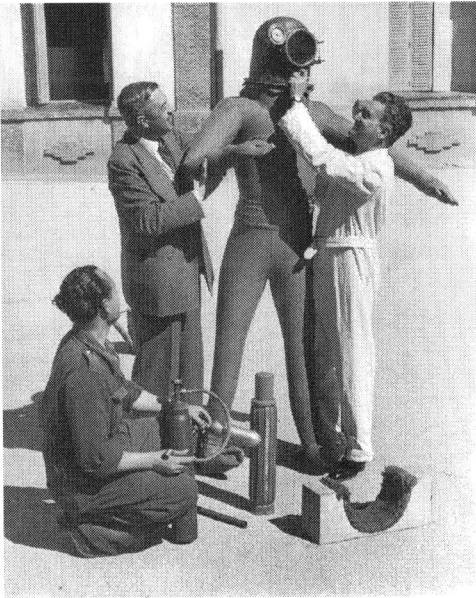

Pressure suits have come a long way since the 1930's. This was a suit designed to be used in a high altitude balloon.

Modern pressure suits are not made of rubber as the hoaxers assume to expand under pressure. Instead all the parts hold in pressure and don't have the property to expand like a balloon. Many materials are capable of doing this. It's that simple!

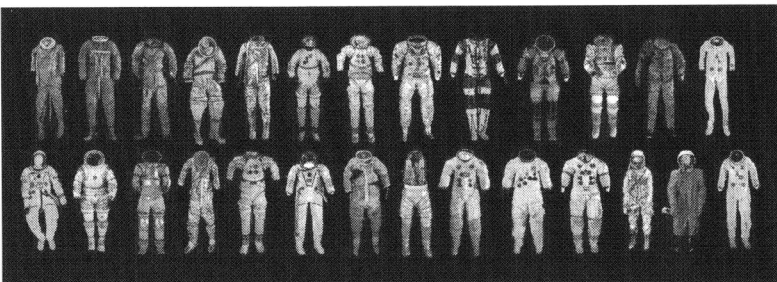

PRESSURE HELMET ASSEMBLY

COMMUNICATIONS CARRIER (SNOOPY CAP)

ENTRANCE SLIDE (ZIPPER) COVER FLAP

SUNGLASSES POCKET

UMBILICAL CONNECTORS:

PLSS O2 OUT

O2 PURGE VALVE

PLSS LIQUID COOLING

WRIST DISCONNECT

PRESSURE GLOVE

HELMET ATTACHING RING

RCU/PLSS STRAPS ATTACHING BRACKET

PENLIGHT POCKET

UMBILICAL CONNECTORS:

COMMUNICATIONS

OPS O2 IN

PLSS O2 IN

CHRONOMETER

PRESSURE GAUGE

PLSS LOWER SUPPORT BRACKET

UTCA FLAP

CDR LEG STRIPES

CHECKLIST POCKET

SCISSORS POCKET

UTILITY POCKET

DATA LIST POCKET

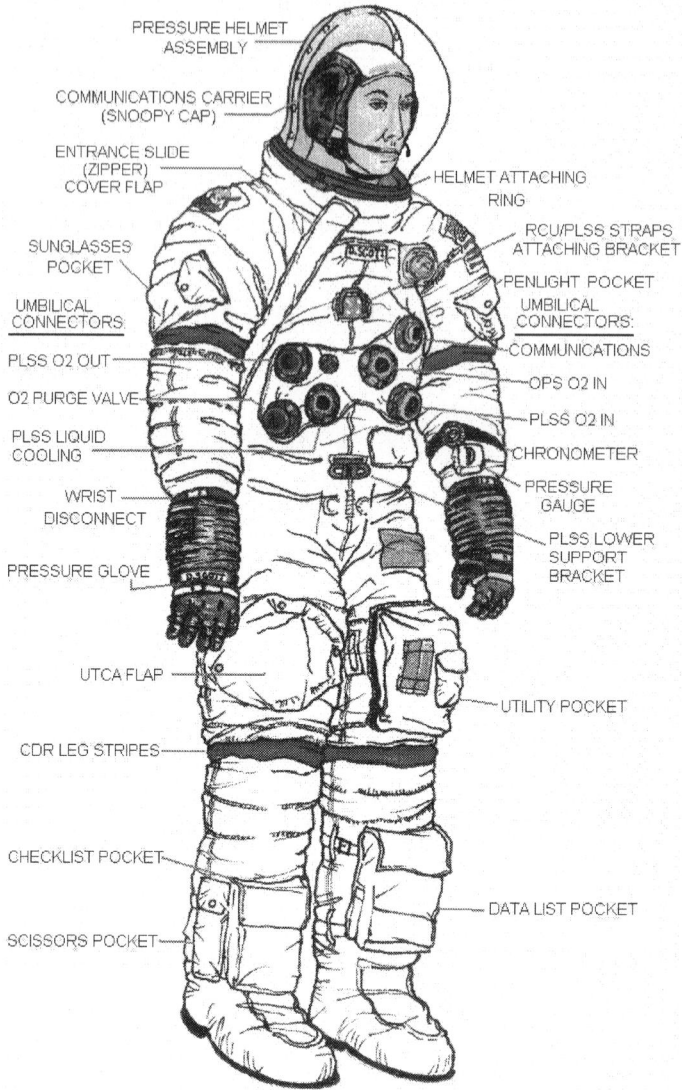

No part of this pressure suit will puff up to such a degree that it prevents the astronaut from working in a vacuum. They are not made of rubber. There are dedicated books on this subject that will explain this in far more detail – one is called 'Space Suits' strangely enough. Purchase one to explore this further.

Lesson learned; modern research is needed; not assuming 'facts' that date back decades.

Chapter 17 Deadly Radiation?

The first US satellite into space was called Explorer 1. It had a simple Geiger counter on board to detect radiation in space. As it was the very first of its kind, nobody knew how many counts of radiation hits would be recorded and so didn't know how to calibrate the equipment. The leading scientist of the time expected around 10-100 counts a minute. The Geiger counter was set for that. After achieving orbit, the hits were higher and went off scale. It could mean 101 hits per minute or 1,000,001.

Explorer 1 satellite being presented to the press in 1958. NASA image.

The next satellite to study it (Explorer 4) would be calibrated better. But in the meantime the press were asking tough questions and wanted answers. The higher figure was constantly quoted for the headlines and let to an impression that space was full of deadly radiation that would kill a human in a few hours.

This radiation zone became known as the Van Allen Radiation Belt after the scientist who interpreted the results. Later satellites proved this not to be the case. Part of the belts did have higher counts but as any spacecraft would only be in it for short periods, it didn't matter too much.

Bart Sebrel on the documentary said 'Hardly anybody knew about the Van Allen Radiation Belts...' Oh really? Pick up any book on astronomy printed from around 1958 and there it all is. It was and still is no secret.

Radiation comes in various forms. Some we are exposed to every day at home, in a hospital, walking around outside. Most forms are completely harmless; it partly depends upon the amount and length of time of exposure. Even a sheet of paper will halt Alpha particles. Our bodies can withstand more dosage than is generally realised. The Hubble Space Telescope, among other satellites, often has its sensors turned off when passing through regions of intense radiation but is rare. An object satellite shielded by 3 mm of aluminium will only receive about 2500 rem (25 Sv) per year, not deadly at all.

Proponents of the Apollo Moon Landing Hoax have argued that space travel to the moon is impossible because the Van Allen radiation would kill or incapacitate an astronaut who made the trip. Van Allen himself who passed away in 2006 had dismissed these ideas. In practice, Apollo astronauts who travelled to the moon spent very little time in the belts and just received a harmless dose. Nevertheless NASA deliberately timed Apollo launches, and used lunar transfer orbits that only skirted the edge of the belt over the equator to minimise the risk.

Astronauts who visited the moon probably have a slightly higher risk of cancer during their lifetimes, but still remain unlikely to become ill because of it. None have developed cancer or anything else related to radiation.

Figures need to be added to this argument rather than just saying 'Space is filled with deadly radiation.' Water is also deadly; a swimming pool can be deadly but not a cup full.

44

Chapter 18 Laser Reflectors

On three of the Apollo landings, 11, 14 & 15, Laser Light
Reflectors were left behind for experiments from observatories
around the world. These were placed on the moon by hand and
aligned with the earth in the sky. As the moon is in a captured
rotation, the earth doesn't move in the moon's sky. Lasers are
fired toward the three sites on the moon and a reflection is
made back to the telescope every time. If you aim the laser at
any other part of the moon, nothing comes back. The reflection
is too weak or even non-existent to record a reflection. Early
attempts at gaining a reflection in 1969 didn't work well as the
laser wasn't powerful enough. It was upgraded and the results
flooded in. Radio waves do indeed bounce off the moon but
light is mostly absorbed.

A list of experiments using the Earth-Moon laser ranging
technique;

Continental Drift rate & direction for earthquake research
Perturbations of up to 1000 asteroids.
Gravitational potential parameters of the Earth and the Moon.
Relativistic description of angular momentum and torque
(Moon).
Relativistic description of torque exerted by the planets.
Newtonian description of non-spherical shape of the Earth and
the Moon.

45

Newtonian description of lunar tides.
Parameterization of Einstein's theory to test some of its
quantities like time variability of the gravitational constant.
Modified Eulerian equations to model the rotation of the Moon
Precession angles of the earth.
Tidal forces due to the oceans and the atmosphere.
Improved description of the non-rigid Earth and Moon.
Einstein-Infeld-Hoffmann equations rule the bodies' movement.
Location of an observer varies due to the Earth tides and the
drifts of the continental plates.
Signal propagation suffers from the Earth's atmosphere
(meteorological data required).
The general gravitational field of the Sun.

Hurstmonceux Observatory, East Sussex, UK. Lasers were fired
from here on a regular basis to laser reflectors on satellites as
well as the moon.

How does it work?

Since the 1950's astronomers at Jodrell Bank Observatory have been bouncing radio signals off the moon to directly measure its distance. It's a simple radar technique. They measured its distance by timing how long it took the echo to come back by using an atomic clock. They divided the total return time in seconds by two, then by the speed of light. The earth & the moon are moving during the experiments, but both motions direction & speed are well understood and taken into consideration.

Formula Echo Timing ÷ 2 x c = Distance (c is speed of light).

Echo from transmission to receive –

2.5644408 seconds ÷ 2 = 1.282204 x 299,792.458 = 384,395.07km to the Moon.

As the wavelength of light is thousands of times shorter than radio, the accuracy with a laser is thousands of times better.

Laser Ranging the moon from the McDonald Observatory, Texas.

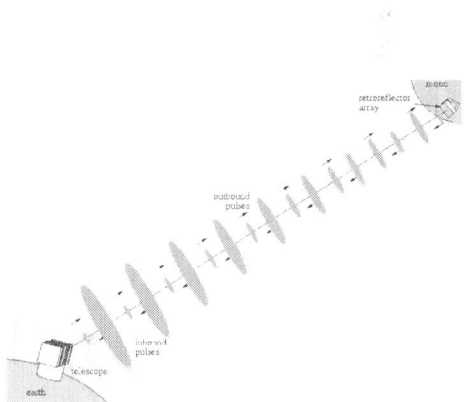

'Well these laser reflectors only acclaim to work via NASA controlled observatories!'

Really? The following observatories fire lasers at the reflectors several times a month. Not one listed is controlled by NASA or even based in the USA. So are all these nations a part of the hoax too?

- TLRS-3, Arequipa Arequipa, Peru 7403
- Fixed System at Balkhash Balkhash, Kazakhstan 1869
- Fixed System at Beijing Beijing, Peoples Republic of China 7249
- Fixed System at Borowiec Borowiec, Poland 7811
- Fixed System at Cagliari Cagliari, Italy 7548
- Fixed System at Changchun Changchun, Peoples Republic of China 7237
- Fixed System at Dunaovcy Dunaovcy, Ukraine 1866
- Fixed System at Evpatoria Evpatoria, Ukraine 1867
- Fixed System at Grasse Grasse, France 7835
- Fixed System at Graz Graz, Austria 7839
- Fixed System at Helwan Helwan, Egypt 7831
- Fixed System at Herstmonceux Royal Greenwich Obs., Great Britain 7840
- Fixed System at Katzively Katzively, Crimea, Ukraine 1893
- Fixed System at Maidanak Maidanak, Uzbekistan 1863
- Fixed System at Maidanak Maidanak, Uzbekistan 1864
- SAO-1, Matera Matera, Italy 7939
- Fixed System at Mendeleevo Mendeleevo, Russia 1870
- Fixed System at Metsahovi Kirkkonummi, Finland 7805
- NLRS, Orroral Orroral Valley, Australia 7843
- Fixed System at Potsdam Potsdam, Germany 7836
- Fixed System at Riga Riga, Latvia 1884
- SALRO, Riyadh Riyadh, Saudi Arabia TBD
- Fixed System at San Fernando San Fernando, Spain 7824
- Fixed System at Santiago de Cuba Santiago de Cuba, Cuba 1953
- Fixed System at Sarapul Sarapul, Russia 1871
- Fixed System at Shanghai Shanghai Obs., Peoples Republic of China 7837
- Fixed System at Simeiz Simeiz, Ukraine 1873
- Fixed System at Simosato Simosato Hydrographic Observatory, Japan
- Fixed System at Tokyo Tokyo, Japan 7308
- WLRS, Wettzell Wettzell, Germany 8834
- Fixed System at Wuhan Wuhan, Peoples Republic of China 7236
- MOBLAS-5, Yarragadee Yarragadee, Australia 7090
- Fixed System at Zimmerwald Bern, Switzerland 7810
- TLRS-1 Mobile system in Europe
- TLRS-2 NASA mobile system in South America
- MTLRS-1 IfAG mobile system in Europe
- MTLRS-2 DELFT mobile system in Europe
- HTLRS Mobile system in Japan

Chapter 19 Perfect Pictures?

Some say that every single picture taken on the moon is perfect; as if taken by a professional photographer. And yet they had their helmets on so it would be impossible to look through the viewfinder. So how can this be?

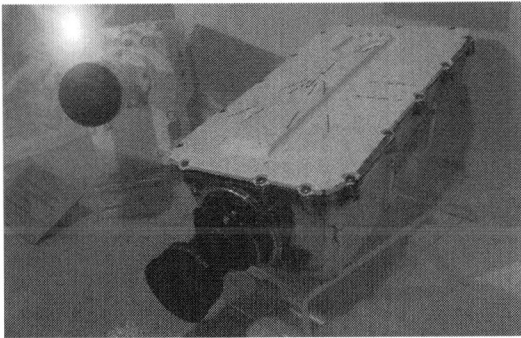

Examples of Westinghouse cameras used during Apollo 11 mission displayed at the Neil Armstrong Museum at Wapakoneta, Ohio. Photos by the author.

The cameras used in and out of the Apollo spacecraft throughout each mission had no view finders. Once a spacesuit helmet is put on, there is no chance of seeing up close what is being filmed. So to save weight the viewers simply weren't added. (LCD screens had not yet been invented).

The simple answer is in several parts. The first three are practice, practice, practice! The training included a course in photography and was given cameras to practice with; identical to the ones that would be used on the moon without a viewfinder. The next answer is very simple too. Not all the pictures were perfect at all. Would you publish in a magazine, book or newspaper any of the following...?

I would fire the editor of any publication if he / she published any of these images. Hundreds like these were taken, errors include out of focus, accidental shutter release before alignment, over / under or over exposure and light bleed at the beginning or end of a roll of film as a film back was being changed; light from the sun contaminating the film ends.

An error in the f-stop setting on the lens caused this under-exposed image. Would you choose to publish this?

As an experienced photographer, I have been in many situations where I had to take shots without looking through a view finder. At a concert for instance; hundreds of heads bopping up and down and I am in the middle of them. The camera is loaded with a known film of perhaps 400 ISO sensitivity. By just seeing the stage I can judge the f/stop and shutter speed required. I may choose 1/125 sec at f2.8. If the stage is more than 100ft away, I can set the focus on infinity. I then hold up the camera in a rough horizontal position, point roughly toward the stage and take several shots at a wider angle than I need to. One image should become acceptable. If the stage is not in the centre or slightly tilted; that can be cropped later. If slightly under or over exposed; that can be improved in a dark room; such a technique was known as 'pushing and pulling.'

A few usable images can be made for publication and yet I never looked through the view finder. Any experienced photographer can do the same. That near perfect image will have been adapted to make it worthy of publication but it doesn't mean the original was fake.

The focusing system was similar to many compact cameras of the time. The wide-angle lens results in a relatively large depth of field. This meant the astronauts only had to get the focusing distance approximately right to get a good sharp image. Instead of a variable focus ring, it was divided into just three preset positions: near, medium and far.

The moon's conditions for photography are also somewhat better suited than on Earth in several ways. As there are no clouds, the sunlight on the moon is at a constant brightness. The surface has a fairly even reflective property; therefore the natural fill-in light from the surface into shadows was also predictable. The spacesuits had a camera attachment clip that ensured a horizontal posture whenever used properly. For the vast majority of the images, the lens can be set at infinity by using marks around the lens. Most of the guess work with taking pictures without a view finder was eliminated. All these points together improved the chances of a taking a reasonably good shot for the history books.

This picture is of my Mamiya 645 camera; a similar model that was used on the moon. I took many perfectly good pictures despite not needing the viewfinder. It does have one but wasn't always practical to use it. It has clear focus settings on the lens and practice / experience allow a photographer to judge the field of view and exposure required. Looking through a view finder is a luxury not a necessity.

The previous image is clearly un-publishable. But a few seconds re-editing the shot and the very obvious error is corrected. This is standard dark-room practice.

I used Photoshop to improve the poorly taken image and produced a publishable picture. It doesn't imply the original was fake. The same adjustments could be made in an old fashioned darkroom.

Lesson learned; *Ask experts within the field in question rather than assume an answer.*

Chapter 20   Who Filmed Armstrong?

**Neil Armstrong, the very first human to climb down a
ladder and walk on the moon. Millions of people saw him
live. But who did the filming then? Someone else must have
been there first. That doesn't make sense. There must have
been a camera operator outside the Lunar Module, it must
have been fake!**

In this age of 'selfies' the answer has to be fairly obvious. A
camera was set up outside on a platform already aligned with
the ladder to transmit the image of Neil and his first steps on the
moon.

*This full size mock-up is at Hutchinson, Kansas; home of the
Apollo 13 spacecraft. Photo by Author.*

A lever was outside the lunar module to open a mechanical mount on the outside; it was within reach as Neil climbed down the ladder. This was a hinged platform with a camera attached. Upon release by Neil as he came down the ladder, the platform sprang out and locked in a pre-set position. The camera was already pointing toward the ladder and Buzz switched it on. Bruce McCandless at Houston announced 'Hey we have a picture on the TV.'

So why does a camera have to have someone looking through its viewfinder to record anything?

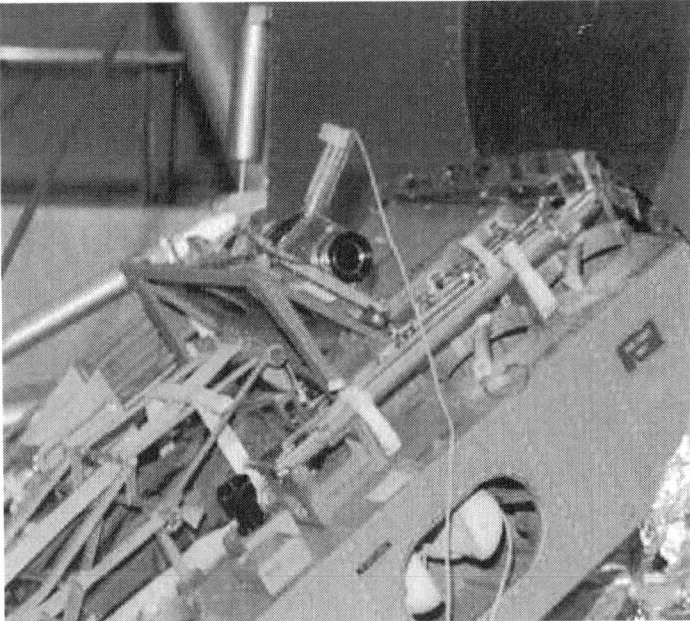

A close up of the TV camera mounted in position aimed at the ladder; pre focused and perfectly angled to record horizontally. The release mechanism was outside the Lunar Module; pull on a bar and this platform unhooked to reveal the camera that automatically switched on.

Any child with a Mechano Set could build something similar. The first remote TV system like this was invented in 1942…

It's not Rocket Science!

Chapter 21 Moving Cameras

Sometimes a camera follows both astronauts and moves to track them across the moon. It even zooms in for a more close up shot. How can this be? Who is doing the filming behind the camera? When the same Apollo craft took off from the moon, it tracked the top stage leaving the moon and flying back up to lunar orbit. Aha; got you all now, a third person as a camera operator was there, it was a TV studio after all!

Anyone heard of CCTV? Remote controlled cameras that record our shopping areas etc for security. There are over 4,000,000 of them in the UK alone. These are remote controlled by someone perhaps hundreds of miles away. There are no people sitting behind the cameras anywhere.

On Apollo 15, 16 & 17, a remote controlled camera was mounted on the Lunar Rover. Designed by NASA engineer Bill Perry, the high gain S-Band radio dish on the Rover transmitted a live feed signal back to the camera controller (Ed Fendell) at Mission Control Center, Houston. He had buttons to pan the camera sideways, up and down, and a button to zoom in and out via stepping motors on the camera and mount. The moon is 1.3 light seconds away so this delay was calculated in and Ed moved the controls slightly ahead of time to record the launch from the moon.

Ed Fendell in Houston at the Camera Console in Houston tracking the launch of Apollo 17 from the moon.

The inventor of the CCTV system was Walter Bruch in 1942, the first was designed and installed in Peenemünde, Germany, to observe the launch of V-2 rockets by remote. It was built by Siemens AG at Test Stand VII. CCTV was used for security for the first time at a store in King's Lynn, Norfolk, England in 1955. Black and white CCTV was used on the moon from 1969 and fully steerable & in colour from 1971.

Apollo 16 Lunar Rover TV camera controlled from Houston.

Chapter 22 Missing Lunar Modules

The conspiracy supporters try to grasp at anything they don't understand in the historic images of Apollo and shout foul play without a moment's thought about what is really happening. If they simply applied a little intelligence and realism they will solve their own questions correctly and halt the embarrassment.

Some pictures were taken of the Lunar Module with a glorious background of mountains, and later another was taken that seemed to show the Lunar Module was 'missing.' Where is it now eh? More evidence of a hoax!

The example here shows the same background; but one has a Lunar Module in it the other doesn't. The film set designers forgot to replace the Module after a cleaning.

Or is it that the astronaut walked to the other side of the Module and photographed the background from there?

When a background is several miles away as with the lunar mountains, anyone can walk in any direction for several hundred yards and the background will hardly change.

Study these above; the background hills and even the clouds. They are same in both images. But the foreground is different with a missing car & author. According to the hoaxers, the car, the author & Monument Valley in Arizona is a hoax. Or is it that his wife walked to the far side of the car or moved a few yards to the left or right in the second picture? Which do you think is more likely?

Chapter 23   Missing Crosshairs

We have discussed missing Lunar Modules, the same applies to missing Lunar Rovers too, but now we have missing 'Crosshairs.'

Crosshairs were etched on a thin piece of glass placed in front of the film plane in the camera for reference points and angular measurements for each picture. They show up as thin shadows on the film. These were known as Reseau plates. The same has been used in Aerial photography for many years. **But on some images the crosshairs seem to go behind objects. They are not in the camera at all; they must be fake pictures!**

Ok let us find something that all such pictures have in common. Every time a part or complete crosshair is 'missing' it always occurs on bright white parts of the picture. Why? Because in photographic terms, that white section is overexposed! No other detail can be recorded. The chemicals in that area have been 'burnt.' The next picture can't even show the photographic grain of the film let alone the cross hair.

Look at the picture above and you will see the point. This is a lens flare from the sun caused by the lens itself. The flare isn't anything physical, so how can these crosses appear behind it? The flare has burnt all the emulsion on the film and cannot record any other feature.

So what are the hoax people really saying about how the crosshairs get on the images if they are not produced in the camera? The crosses must have been hung on wires; all aligned perfectly a set distance apart throughout the shot. When the camera moved a few degrees to take the next picture, all the cables would have to move too. What is the point? Why not just leave them out or better still etch them on a piece of glass and put in the camera to cast a shadow on the film; exactly as they did?

61

The crosshair on the flag is part missing on the white stripes only. But the hoax guys say that part of the cross is behind the flag the rest is in front.

Do you see how daft the whole idea sounds?

The crosshairs were placed on such images within the camera as reference points. Scientists can talk about various rocks or mountains and refer to a picture number and a grid reference on each shot based on the crosshairs or take angular measurements between mountains etc.

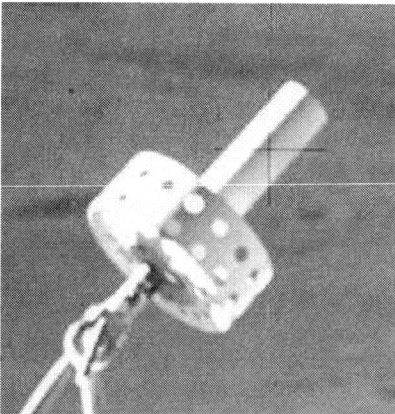

from NASA Frame AS17-146-22296

How is a part of this crosshair supposed to go behind this instrument? Or does it just seem to be going behind but it's really burnt out due to the intense white light?

I just wish conspiracy people would just stop and realise where they are going wrong. I will make scientists out of them yet.

62

Here is the full answer... This is a close-up of my own camera. It is similar to the models used on the moon. Look carefully into the waist level finder and there is a grid screen. It is fully visible when taking pictures to align the horizon and adjust for proportions of landscape and sky etc. In this case the grid is not on the film plane and did not project onto the pictures; but the Apollo cameras were. The grid wasn't going behind anything; the crosses just simply cast a shadow on the film.

Wherever the crosses or anything else appeared on intense pure white parts of an image, that part was burnt out and could not be recorded as the chemical emulsion on all layers of the film base was used up. Ask any professional photographer.

Chapter 24 Spotlights turning on?

What is going on in these pictures from Apollo 11? In one shot the Lunar Module is in deep shadow and it can hardly be seen. The next shows it brightly lit. I know what it is... A spotlight! Aha got you all now. Yes it's in a studio after all!

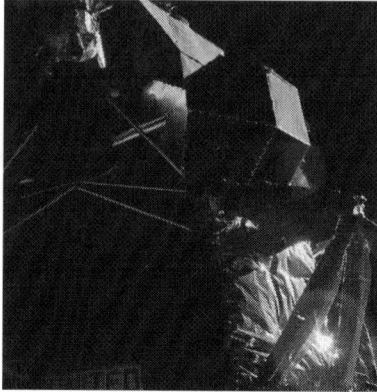

This particular photo above by Neil Armstrong was taken at a smaller f-stop than needed as this side was in shadow. The lunar surface itself reflects light into shadowy areas. Neil then opened the f-stop ring to allow more light onto the film. Images of Buzz Aldrin coming down the ladder were then perfectly exposed. It had nothing to do with spotlights. Ask a photographer!

Two pictures showing the lens f-stop aperture. This can be
manually controlled by turning a ring around the edge of the

65

lens. As the astronauts had thick gloves on, they had a 20mm long metal bar protruding from the corresponding ring to allow manual setting instead. They were told that for all the lunar surface pictures, set the exposure speed to 1/250 second at f11 for sun exposed areas, and f5.6 for the shade.

This is one of the cameras used by Neil Armstrong. Note the 'arm' welded onto the aperture ring. This allows an astronaut to adjust the aperture of the lens even with thick gloves on. Image taken by the author at the Neil Armstrong Air & Space Museum in Wapakoneta, Ohio.

It is common practice to take pictures that are under difficult lighting conditions (part shade / part sun) with slightly different exposures to ensure at least one perfect shot. All photographers do this. This technique is known as 'bracketing.' The astronauts were also trained to do just that. The previous lens pictures show the aperture partly closed for the corresponding darker image of the Neil Armstrong museum and wider for the lighter. Ask ANY photographer about this. A wider aperture will allow more light on the film, a brighter image. Modern digital cameras often have a facility that takes three pictures every time a shutter button is pressed. One slightly under exposing, one about right, and the other slightly overexposed. That's all folks!

Above; The Neil Armstrong Air & Space Museum in Wapakoneta, Ohio. The first picture was underexposed by around 4 f-stops. The aperture was widened for the second picture and was correctly exposed. Or did I switch on a spotlight in a massive studio?

The hoax supporters need to study photography rather than guess. They never do and are fundamentally lazy. If they attended classes, a course or just read some books on it, they will find the correct answer to lighting and exposure.

Chapter 25   No engine crater!

Early artist impressions of what a Lunar Module would look like on the moon showed a crater under the engine. It was assumed by the artist that the thrust from the engine would produce a deep gouge in the lunar surface.

The artist was great at art, but not on physics & geology. The crater shown above is not correct.

It may at first seem obvious that this would be the case. But if the actual thrust per square cm from the engine onto the surface was calculated and the toughness of the moon was taken into account, then a different result occurs. (A crater forms when a solid chunk of rock hits the moon at 10 km/s or more. Gas from an engine is not the same intense force). *NASA image below.*

This Apollo 12 image shows a shallow depression where dust was blown away but not a deep crater.

The hoax believers claim the LM descent stage used its full thrust of 10,000 pounds at lunar landing and that it should excavate a large blast crater under the LM. At landing in the low lunar gravity (which is 1/6 of Earth's gravity), the LM throttled down to about 3,000 pounds of thrust. The LM had 6 foot long landing probes under 3 of the 4 footpads and when any of the probes contacted the surface, the crew shut down the engine so that it would fall the last few feet to the surface; the engine was more than 6 feet above the surface when it stopped firing.

The dust is clearly visible flying out at high speed away from the LM prior to touchdown in all of the lunar landing films taken from the LM cabin windows during approach and landing. Given that the descent stage engine bell is about 5 feet across at the bottom, and that thrust of the engine at touchdown was about 3,000 pounds, the blast pressure of the rocket exhaust was only about 1 pound per square inch; less power than a firework.

69

V2 rockets used to take off with just dirt ground underneath, but no blast craters. Why would we expect to find a blast crater under the LM? It certainly blows away some of the surface dirt in a radial direction and will create a small depression, but not a crater in concrete or similar. There is even an Earthly example of a rocket landing on dirt. The DC-X was a test flight program of a vertical takeoff and landing rocket. On one of its last flights, it made an emergency landing outside of the pad area. Despite the hydrogen/oxygen engine producing a thrust of some 60,000 pounds, the engine produced a mark on the desert floor that was barely recognizable.

Look at dirt in an area where a powerful firework has taken off... see a crater?

No rocket launch like this V2 ever leaves a 'blast crater'.
Scorch marks perhaps, sand blown away yes but not a crater.

Chapter 26   The Film Would Melt!

When we announce temperature on Earth, we normally measure air temperature. Since there is no air on the moon, we can only measure or take into account surface temperature instead. On a hot day, even in the UK, the surface of a car can reach 80° C or more depending upon the colour. A camera may be left outside in the same conditions, but have you heard of film actually melting inside? I have done the same on several occasions but nothing happened to the film. Upon developing it, one single frame may have been exposed to light that has bled into the back of the camera and was ruined. But the rest was fine.

Ever since astronaut Wally Schirra brought along his own Hasselblad 500C on his Mercury mission in 1962, NASA chose Hasselblad to make cameras suitable for future space missions. It was not until NASA saw the quality of Schirra's photos, that they realized the importance of documenting their missions with quality photographs.

The day on the moon lasts for approximately fourteen earth days. There is more time to heat the ground; anything else brought by the astronauts is only there for a much shorter period of time. Plus objects that are moved around such as cameras are sometimes in the shade, or changed orientation with the sun. The temperature of such objects will be much more evened out. The surface temperature of the moon's surface near the equator at midday is approximately 123° C. This is due to it being stationary and sunlight from directly overhead. Any other objects would have a lower temperature.

These facts were known long before Neil & Buzz walked on the moon. To ensure the film wouldn't be affected too much by the heat, the film used was Kodak Panatomic-X fine-grained, 80 ASA black and white film and Kodak Ektachrome SO-68. These were chosen partly for their resistance to heat; often used in desert conditions on Earth. The cameras were also silver in colour to reflect much of the heat without using any energy.

The film was contained preloaded in 'Bulk-Film-Backs.' They were simple cube shaped compartments that contained a pre-loaded extra large role of film. The same system was used by some Mamiya medium and large format cameras. My last film camera was the Mamiya 645 Super which did incorporate bulk-film-backs. This is to save time during any kind of rapid photo shoot such as a wedding (I have done many in my younger days as a pro-photographer)). Changing a film-back can be made in around 20 seconds instead of several minutes and the photographs can commence again, plus 120 pictures or so can be taken instead of 16 before changing the film.

Astronauts on the moon not only had to work fast but also had the handicap of those thick gloves while out on the surface. The camera-back design was perfect for lunar photography. The only disadvantage is the potential of light leaking into the camera around the film-back attachment area during the change – shown on the previous page. So the astronauts were told to take the first picture of each roll of film twice or just use the first image on anything.

Chapter 27   Six mile long spaceship?

The picture below is obviously the shadow of the Command
Module (CM) of Apollo 11 on the Moon itself. The trouble is
that crater to the bottom right is around 1 mile wide, that
makes the length of the CM around 6 miles long. Got you
NASA!'

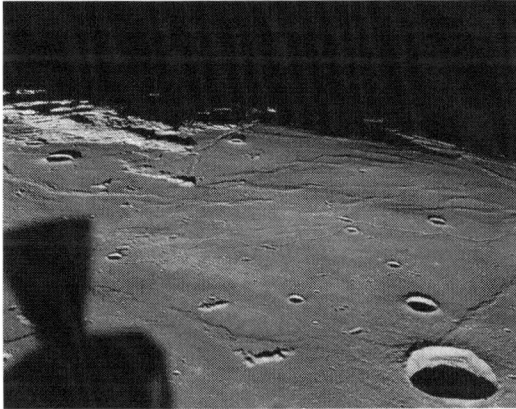

*What Rot! That picture above has been cropped. See the full
version below. Spot the part that was cropped out?
Crafty aren't they?*

It is not a shadow of the Command Module at all. It is a silhouette of the Vernier Engines just outside the Lunar Module window. The same is on the opposite side of the LM.

I have used Photoshop to enhance the image and show more detail of the thrusters to prove the point. The round shape of the nozzle is now clear and cannot possibly be a shadow.

Check out another picture from lunar orbit from the same roll of film and compare it to the shape of the thrusters as seen in the Air & Space Museum, Washington DC.

Lunar Module thrusters on a full size trainer on display at the Smithsonian Museum, Washington DC. Below is another at Huntsville, Alabama. Recognise them? Images by the author.

Lesson learned; Don't take their word of what you are seeing! Examine the original evidence, not the tampered version by the conspiracy supporters. They will do anything to convince you they are right.

Chapter 28   Giant Earth!

'As shown below, the Earth is FOUR times wider than the
Moon. Therefore the Earth should look massive in all the
photographs from the moon. It doesn't, it still looks small.
So all these pictures are out of scale; more proof of a hoax!'

Is this conclusion correct? Nope! The lens used most often on
the moon was a wide standard lens without magnification
unlike a zoom or telephoto. From the Earth the moon appears
just 0.5° wide. From the Moon, the Earth appears 2° wide.
Photos taken with a standard lens are around 80° across. Hence
the width of the Earth will be just 1/40th of the picture's width
across. Would it look huge? Nope!

It is so simple to discount such claims and yet not one of them
ever bothers to check their own assumptions with figures.

This above NASA image has been cropped from the full frame.
So the original full picture will show the Earth even smaller, but
this was quite normal. Just a couple of simple calculations
prove that point beyond doubt.

In the famous 'Earthrise' photo by Apollo 8, the earth looks far
bigger now even though they were still close to the moon.
That's because the camera had interchangeable lenses and a
300mm telephoto lens was now used instead of the standard
lens used on the moon.

Chapter 29 Giant Sun!

Well we have had a supposed Giant Earth, guess what the Conspiracy guys now have thought of? You may have guessed it, there is a Giant Sun... Which can't be true so it must be a spotlight instead. We are off again...

This is a Gigantic Sun, totally out of scale; it's a spotlight!

Ah! Houston we have a problem!

This picture is perfectly normal - ask any professional photographer. Do note that the light is coming from beyond the Lunar Module and not just Alan Bean on Apollo 12. How big would that make the spotlight then? 50 metres wide?

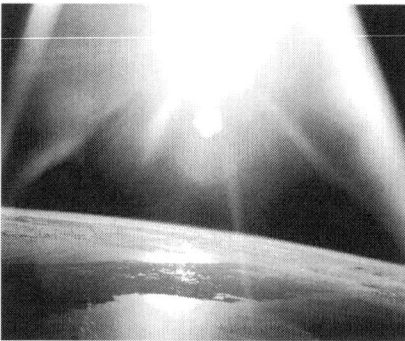

An Apollo 7 image from Earth orbit. The sun looks huge compared to how we see it so it can't be real can it?

What is going on? Well in photographic terms the sun in this and other similar pictures is totally overexposed. The camera was correctly exposed for the moon's surface or the earth itself, not the sun. The build up of intense light has created a massive lens flare around the sun itself as well as extra flares within the lens. But pictures of the sun on Earth can have exactly the same effect.

According to the Hoaxers, Bodium Castle that has stood there for over 1000 years in Kent is a fake! The image of the sun is far too big and out-of-scale; it must be a spotlight on a model sitting in a studio.

Or is it that the sun is too bright for the camera to handle with the setting I gave it? You decide which is more realistic.

I honestly wish that these conspiracy guys just pop along to a local photographic club and ask such questions. They will get the right answers first time. Better still; they should take their own pictures of such conditions. They will never bother though, but keep bashing on about subjects they know little about and take their totally pure guesses as fact.

This is me in a glider over southern California. I'm really showing off now. Is the glider and California fake because the sun looks too big? Or is the sun over-exposed? The same optical effect occurs with digital as well as film, on earth as well as the moon.

And who is taking the picture? It really must be fake after all! I gather you can work that one out yourself.

Lesson Learned; Reproduce similar conditions with similar equipment. Observe the results, accept them and don't guess.

Chapter 30 Silent Engines

Bill Kaysing on 'Did they go to the Moon?' claimed that you should hear the sound of the descent engine on the Lunar Module in the audio from the landings. There are several problems with this hypothesis;

The engine is many feet away in a vacuum so that the sounds would have to be transmitted through the spacecraft structure itself; sound cannot travel through a vacuum. I do realise that launches from Earth are within the atmosphere; however the complete answer lies in the microphone itself.

The microphones used are often insulated inside of the spacesuits worn by the astronauts especially during launch and other crucial phases of the mission. The situation occurred on Space Shuttle launches. Are they saying all 118 shuttle flights were faked too just because we can't hear the rocket engines in the voice communications? Not only the US launches but what about the Russian and now Chinese launches. Are they all fake too now?

Bill Anders of Apollo 8, December 1968. They all used microphones that restrict the direction of the sound source so voice is heard only. This is sometimes referred to as 'VOX' or 'Hotmike.'

The microphones are worn next to the astronauts mouth and are designed only to pick up sound from its immediate vicinity and are background noise-cancelling by design. They are of the same type used by pilots of all kinds of aircraft. When a pilot radios back to a Radar Controller as seen on documentaries, do you hear the sound of the jet engines in the background? No!

Racing car drivers have the same system, you can't hear the screaming engines on the radio communications from the drivers either, just the voice. So perhaps Formula 1 is all a big hoax too? Such microphones are known as 'uni-directional.' They are designed to pick up a specific sound range within a narrow beam width. Any other sound sources are not picked up by the microphone; hence engine sounds from the side etc would not be transmitted.

The noise an astronaut does hear from a rocket engine while in space is created in the boundary between the exhaust and the ambient air. If there is no ambient air (in space) only internal combustion and flow noise can be heard; hence a very limited volume of sound. (A contribution by Allan Folmersen).

There are many similar examples in other trades using uni-directional microphones. These hoax supporters must stop guessing and deal with the real world!

Sally Ride with a uni-directional VOX microphone. Racing car drivers use the same, but no high pitched screaming sound from the engine is heard coming through the transmissions.

82

Chapter 31   Flameless Engines

The top half of the Lunar Module launched from the Moon to bring the astronauts back into orbit, there was no flame from the engine. Obviously, every rocket has a visible flame, so the take off was faked... caught you again NASA!

There is a simple reason why you cannot see the flame from the Lander when it took off. The fuels they used produced no visible flame! The Lander used a mix of hydrazine and dinitrogen tetroxide (the oxidizer). These two chemicals ignite upon contact for reliability and produce a flame that is transparent. We expect to see a flame because of the usual drama of lift-off from the Earth; the flame and smoke we see from the Shuttle, for example, is because the solid rocket boosters do actually produce them, while the lunar Lander did not. The fuel mix was completely different; plus the burn took place in a vacuum, our atmosphere contains dust and pollen that burns, the pad is also doused with thousands of gallons of water; launching from the moon did not have such properties.

The British rocket called Black Arrow used Hydrogen Peroxide (HTP) and Kerosene as fuel. When burnt, hardly any flame was seen. The fuel mix, water on the pad, moisture and dust etc in the air, or launching in a vacuum are all variables that have to be taken into account.

A flameless engine launching the British satellite Prospero into orbit in 1971.

Rocket engines do not have to have exciting visible flames shooting out. This is due to fuel mixture and partly due to air moisture. The dry air in Australia helped with this effect. Moisture on the moon or in space is zero.

Chapter 32   Rover too BIG!

"Look at the size of that Rover, how did it fit in that tiny Lunar Module?"

The hoax supporters claim that the Lunar Rover was too large to fit in the Lunar Module. If one takes the measurements of the Lunar Rover Vehicle (LRV) when it was fully deployed and assembled, then yes, it would not fit in the Lunar Module. However the Rover was folded for stowage in the descent stage of the LM in a quadrant to the right of the ladder. The chassis was hinged in three sections and the four wheels were pivoted nearly flat against the folded chassis occupying only 30 cubic ft. When the astronauts deployed the Lunar Rover, all they had to do was pull on two cords and the Rover popped right out of its berth and down to the lunar surface.

LRV DEPLOYMENT SEQUENCE

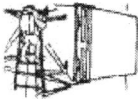

- LRV STOWED IN QUADRANT
- ASTRONAUT REMOTELY INITIATES AND EXECUTES DEPLOYMENT

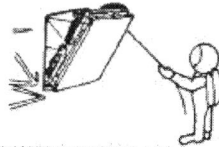

- ASTRONAUT LOWERS LRV FROM STORAGE BAY WITH FIRST REEL

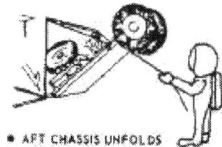

- AFT CHASSIS UNFOLDS
- REAR WHEELS UNFOLD
- AFT CHASSIS LOCKS IN POSITION

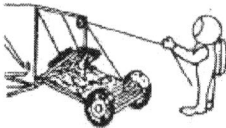

- FORWARD CHASSIS UNFOLDS
- FRONT WHEELS UNFOLD

- FORWARD CHASSIS LOCKS IN POSITION. ASTRONAUT LOWERS LRV TO SURFACE WITH SECOND REEL

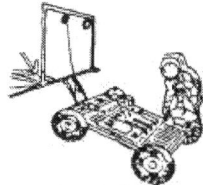

- ASTRONAUT UNFOLDS SEATS, FOOTRESTS, ETC. (FINAL STOP)

All four wheels folded in, the seats folded down, then the whole kit was folded backward in two places on large pin/hinges. Full details are given in a Haynes Manual on the Lunar Rover available from www.amazon.com. A direct link is on the website.

84

As far as the weight is concerned, the moon's gravity is just 1/6th, pulling it out of the compartment and unfolding it would be no problem. *Above photo by the author. Below; NASA.*

I just wish these hoax people will stop guessing at matters like this one. It is so easy to get the truth, but they won't visit a space museum to see back-up artefacts or full size replicas. The mechanisms aren't military secrets; it's just advanced Mechano that any serious engineering student can design.

85

Chapter 33   Where did they go?

Ask any hoax supporter this question...

"If the astronauts didn't go to the moon then where did they go instead?"

The usual reply is that they reached earth orbit after launch and stayed there. **"The moon trip was fake but not the orbiting bit."** The problem with this is that anybody can just look up at the sky and see satellites passing over. Try it yourself, they look like moving stars. The Apollo command module, service module and lunar module together were a substantial size. It would have been the brightest satellite of the time. You couldn't miss it. I have written an entire book and website on the subject 'Satellite Spotting & Operations Handbook' available via **www.satellitespotting.co.uk.**

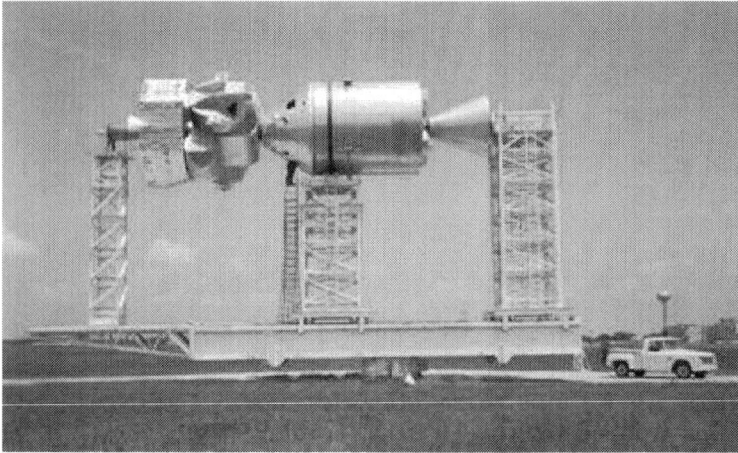

This vehicle was huge. In earth orbit, it would have outshone every star in the sky. It was never seen once it left the earth by any of the hundreds of thousands of people who gaze at the night sky regularly. It should have been spotted passing over every orbit during Apollo 8, 10, 11, 12, 13, 14, 15, 16 & 17. So why weren't they seen? They were on their way to the Moon!

Observations were made of Apollo 7 & 9 passing over for the duration of those missions as they did indeed remain in Earth orbit. Apollo 7 tested the Command Module and then Apollo 9 concentrated on the Lunar Module with engine burns and docking / undocking practice etc.

I have recorded many satellites with an image intensifier and even a standard camcorder. It just demonstrates how easy they are to see. The Apollo-LM configuration would have stood out like a beacon; brighter than any other satellite at the time.

I recorded this satellite above passing over my home town. It was SkyMed 2 and is only a little larger than a car. The shutter was left open for around 20 seconds. As the satellite passed through the field, it was recorded as a streak. It faded as it passed into the Earth's shadow.

The Apollo configuration was much larger. So large in fact that anyone with a telescope using a magnification of 50 or more would have just seen its shape and knew what it was. This was never reported by the thousands of amateur astronomers around the world as they were on their way to the moon instead; too far to be observed in any great detail.

We also need to remember that amateur radio hams also tracked the Apollo missions. The frequencies used were not national secrets. With the right equipment and know-how, thousands of radio hams tuned in to the transmissions. Once the trans-lunar burn was made, the transmissions began to fade and eventually became too weak for them to be heard until they returned from the moon.

Even today, radio hams listen in to the transmissions from the International Space Station as it passes over. On occasions, an astronaut onboard will volunteer take on the role of setting up two way chatter with radio hams. They have to be brief as the pass only lasts around five minutes, then the station passes over the horizon.

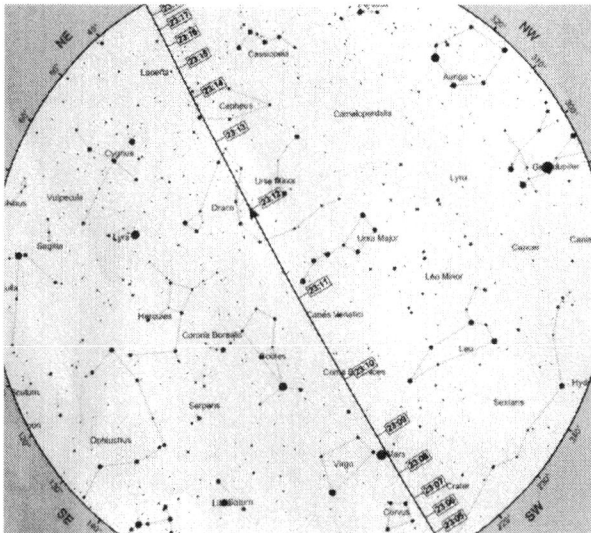

Above; all low orbiting satellites pass over in just five minutes or so in relation to a fixed point on the ground.

Chapter 34   Flying Bedsteads

Lunar Landing Research Vehicles (LLRV)

These strange looking contraptions were Lunar Module (LM) trainers. All the LM pilots practiced in these things. Four versions were made. They handled in a similar fashion as a real landing on the moon. Each pilot flew in these dozens of times.

But on one particular day, a few months before the Apollo 11 flight, Neil Armstrong flew an LLRV and a thruster on one side got stuck on. The craft tilted over and Neil ejected just in time and parachuted down. The Flying Bedstead as it was known, turned over and crashed. Neil's timing of the ejection was perfect and he was cool about the whole incident after. It was this accident that almost certainly decided that he was going to be on the first moon landing attempt.

The hoax supporters keep bashing on about this particular accident. They say that "if a skilled pilot can't land this on Earth, how is he supposed to land one on the Moon?" They completely ignore the 344 other successful flights of the Bedsteads and always quote this one disaster. Besides, the thruster that failed didn't even exist on the real LM, so the same problem wouldn't occur on the moon anyway.

Above; a Surviving LLRV at the Virginia Air & Space Center, Hampton, Virginia • Built in 1965, the simulator is a manned rocket-powered vehicle used to familiarize the Apollo astronauts with the handling characteristics of a lunar-landing vehicle. This particular one was controlled on wires; others had jets and were free from restraints. Photo by the author in 2008.

Lesson learned; to get to the truth, we have to investigate all the statistics, not just a chosen few or even one.

Chapter 35   Washing Machine Computers

The hoax supporters often mention that the computers on the Apollo spacecraft were no more powerful than the processing boards inside a standard modern washing machine. This is true...

But unlike general-purpose computers today, the Apollo guidance computer had to perform only one task at a time - guidance or engine burn length etc. Most of the number crunching was performed at Mission Control mainframe computers at the Johnson Space Center, Houston. The results were then transmitted to the onboard computer, which acted upon them. The Apollo guidance computer was capable of computing only a small number of navigation problems itself. Since the guidance computer had to run only one program, that program could be put in ROM, thus only a small amount of RAM was required to hold the temporary results of guidance calculations. Once the data was used, the memory was cleared for the next phase.

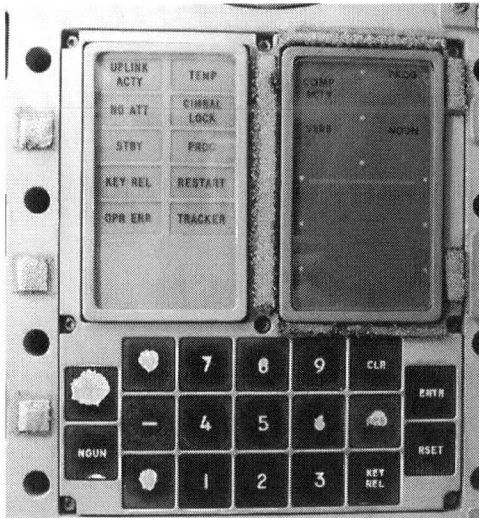

The hoax supporters tend to overrate the tasks performed by the onboard Apollo guidance computers of the 1960's. In fact, the Mercury spacecraft, 1961-63, flew into space without any onboard computer whatsoever; the early Soviet craft didn't either. Yet the trajectories were precisely controlled and the capsule was capable of fully automated & manual control. A picture of the Apollo computers from the Command Module is shown previous. Note; these computers did not require a 1GB windows system either.

Above; Mission Control Center, Houston, Texas. All the serious calculations and monitoring for Apollo was carried out on the ground. The Apollo spacecraft only needed a few numbers on board to control it. Even space missions today follow the same format. Photo by the author in 2011.

Some hoax supporters even go as far as to say that programmable computers were not even invented during the Apollo era. The first such electronic machine was built in the UK in 1943 called Colossus and was used to crack the German Enigma codes. It even used paper tape as NASA later did over 20 yrs later.

Chapter 36   Area 51

All kinds of top secret projects were performed in Area 51; a very restricted base that only a selected few military personnel have access too. This is located north west of Las Vegas and has chartered planes taking staff direct from the McCarran Airport to the base. There even craters there produced by atomic explosions and astronauts did indeed use for training. So it is natural to assume that the moon hoax filming took place there too.

Nuclear bomb craters at Yucca flats near the Area 51 base in Nevada. USAF image.

The Lunar Rover was tested here as the terrain somewhat matched lunar a lunar landscape. Plus the astronauts gained experience on studying such features without venturing down the slopes. On the moon, slopes may seem shallower than they actually are due to the lighting and soil brightness. Without such awareness an astronaut could easily misjudge and cause injury.

It has been claimed that such training was kept secret as the public would make an obvious link between this location and the 'fake' moon images. So how come these pictures displayed here plus more were published in Life Magazine during the 1960' & 70's? Such training was always in the public domain and was never a secret.

NASA images

Chapter 37   Not been back!

Since the final moon landing of December 1972, no astronaut has set foot upon the moon. Why not? We can't get an astronaut on the moon today, the equipment doesn't exist. Aha! If we can't do it now with 21st Century technology then how could the USA possibly do it in the 1960's and 1970's? It must have been a hoax!

Anybody recognise the aircraft below? You got it - Concorde. It's first flight was in 1969, final flight took place in September 2006. We now cannot fly 100 passengers at Mach 2 (twice the speed of sound) in any aircraft anywhere on Earth. Therefore all 16 Concords built, all passengers & crew that flew on them for 37 years must have been lying. Concorde was a hoax! You agree with this argument?

If you don't, congratulations! Silly isn't it? Concorde became too expensive to operate. All kinds of other projects now don't operate for various reasons; Hydrogen airships and Space Shuttles for instance.

We don't use Hydrogen Airships today; does that mean this was a hoax? Or was it because that after two major disasters they were considered as too dangerous?

Have these pilots as well as all the thousands of passengers for 37 years been lying about travelling at Mach 2 as it can't be done today? Or is it because the flights were becoming too expensive and risky to continue after the Paris air disaster?

The moon landings came to an end as the Vietnam War became too costly and the risk of loosing astronauts on a moon flight was too high to continue especially after the experience of Apollo 13. NASA's luck would have run out at some point. It was decided to move on to space station technology for a time before heading back for the moon. China may well be next.

96

Chapter 38   Lack of Technology?

"They couldn't have possibly put people on the moon in the 1960's. The technology wasn't good enough then. We only had black & white TV, no solar power, no computers, not even calculators!"

How true is this statement? This is a common point often mentioned on You-Tube. But let's look at real technological history related to Astronomy and Spaceflight rather than guessing yet again.

1800; Discovery of Infra-red, William Herschel, UK.

1835; First electrical Solar Power, France.

1905; Discovery of time travel into the future, Albert Einstein.

1910; First Particle Accelerator at Cavendish Laboratory, Cambridge, UK.

1917; The atom was split in half at the Rutherford Laboratory, Cambridge, UK.

1938; Baird produced world's first experimental colour TV broadcast, London, UK.

1942; CCTV invented for V2 rocket observations, Peenemünde, Germany.

1942; First Nuclear Reactor built in Chicago, USA.

1943; First Programmable Computer designed by Alan Turing at Bletchley Park, UK.

1944; First Rocket into Space, the V2, Germany.

1944; First jet powered fighter aircraft, ME 262, Germany.

1945; First Atomic Bomb, New Mexico, USA.

1947; First plane to break the Sound Barrier, California, USA.

1947; First solid state solar cell, Bells Labs, New Jersey, USA.

1954; First Nuclear Power Station, Obninsk, USSR.

1957; First electronic calculator by Casio, Japan.

1957; First satellite, launched by USSR.

1959; First spacecraft to fly past the moon, USSR.

1959; First mass produced Silicon Chip, invented by Robert Noyce, California, USA.

1961; First Man in Space, USSR.

1962; First spacecraft to reach Venus, Mariner 2, launched from Florida, USA.

1964; First spacecraft to reach Mars, Mariner 4, launched from Florida, USA..

1966; First regular colour TV broadcast - BBC, London, UK.

1966; First Vertical Take-Off Jet to enter service - The Harrier, UK.

1969; First supersonic & fly-by-wire passenger aircraft, Mach 2 - Concorde, UK / France.

So did we have the technology to go off to the moon in 1969? You bet we did - but only just!

Chapter 39   The Lost Blueprints

"The blueprints for the Saturn V rocket has been lost or destroyed."

It is unsure as to why this claim is supposed to imply a hoax. If the hoax supporters mean that there is no complete set of blueprints in one place, then yes, this does not exist, nor has it ever. The millions of documents relating to the Saturn V and its components were spread out across the country among several NASA centres and thousands of contractors. Certainly many copies of these documents have been discarded, but much of it still exists. Marshall Space Flight Center in Huntsville, Alabama has such documentation on microfilm and the Federal Archives in East Point, Georgia has 2,900 cubic feet of Saturn documents. Rocketdyne, who built the F-1 and J-2 engines, has in its archives dozens of volumes from its Knowledge Retention Program.

"Plans for the Saturn V, Lunar Module and Lunar Rover have been destroyed and no longer exist."

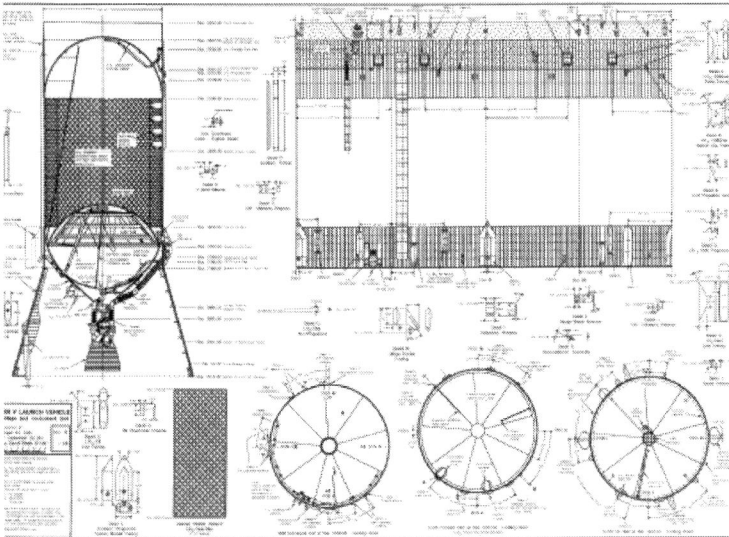

Much paperwork relating to the Lunar Module and Rover has been discarded; however this is to be expected. No company is going to keep in storage millions of documents for an obsolete project that has no chance of being resurrected. But it is not true to say the documents no longer exist. The National Archives microfilmed everything they thought was historically significant and those films are currently in storage. It is not uncommon for space enthusiasts and modellers to find many obscure facts and details about the LM, Rover, and other Apollo hardware from this archive.

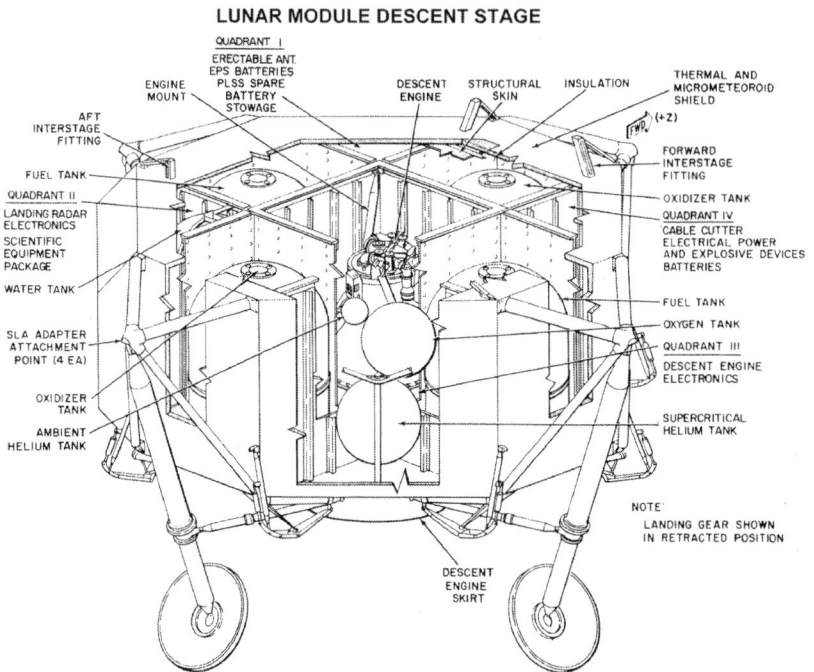

LUNAR MODULE DESCENT STAGE

Hundreds of thousands of documents like these still exist in the form of microfilm or digital images!

A complete set of blueprints of the world's first Particle Accelerator don't survive to this day nor does the very first aircraft, HMS Victory or even the Titanic. Does it mean that they didn't exist? It's another ridiculous claim that they just don't research but just assume.

Chapter 40 The Cold War Connection

Russia was ahead in the Space Race for a decade. American citizens were naturally concerned that it was possible the USSR could put nuclear weapons into orbit and drop them down in minutes, but the USA didn't have the same potential capability to even the score. President John F. Kennedy, after taking office in 1960 was also deeply concerned about the issue. The Cuban Missile Crisis of 1962 demonstrated how close both sides were to a full nuclear conflict that could kill hundreds of millions of people within hours. It would have made WW2 seem like a gang fight. A way had to be found to improve the US technology and show the USSR what they are capable of without detonating nuclear weapons in anger or putting them into space. JFK announced their intention of placing a man on the moon by the end of the 1960's; the perfect answer.

Once Armstrong & Aldrin had touched the moon in 1969, the USSR realised how much US technology had improved. The power & accuracy of the new US missiles were recognised and the Mutually Assured Destruction (MAD) scenario had formed. Regardless of who fired first, both sides would loose.

1975 saw the first joint space mission between the US & USSR. A handshake took place in Earth orbit above Bognor Regis, England UK. Arms limitation talks began, followed a few years later by arms reduction till the fall of the Berlin Wall. The peaceful space program did much to allow this process to take place; a process few people appreciate to this day and just complain about its expense instead.

Deke Slayton & Alexey Leonov commanded the first joint US / Soviet space mission. This marked the beginning of the end of the Cold War.

Chapter 41   The Spies

Some 400,000 people worked on Project Apollo. Some spies from the former Soviet Union did penetrate NASA during this era. Most were known, the CIA preferred to keep track of known spies rather than reveal them, arrest them, only to be replaced by new unknown spies. Information did leak out to Russia regarding the Apollo technology, but nothing was ever found and exposed by them then or since the Cold War ended.

The most secret science project the USA had ever performed was the Manhattan Project; the development of the Atomic Bomb in WW2. After the war, the CIA discovered several spies penetrated the program; Julius & Ethel Rosenberg, Klaus Fuchs and David Greenglass. The USSR built their own Atom bomb and detonated it via information from just four spies. If they were that good at spying, then why couldn't they discover the moon landings were a hoax? The answer is very simple... It wasn't a hoax!

Above; The Los Alamos site where the atom bomb was built. Now a park for everyone to enjoy. Photo by the author, 2014.

Russia's first Atomic explosion 29th Aug 1949. Based largely on the US design that helped to end WW2. The Manhattan Project wasn't as secret as at first thought. Within four years, the Soviet Union built their own based on the plans leaked out by just four people. It would only have taken one single spy in the right place amongst the 400,000 staff to reveal the Moon Landing Hoax. It never happened.

Off coarse the hoax guys say that "Well they had secrets of their own that the USA knew about so they did a deal for both sides to keep secret." Bill Kaysing.

Both sides were spending Billions of Dollars / Roubles trying to gain something on the other risking a global nuclear conflict as they do... then they do a deal to avoid an embarrassment? Why hasn't the new Russian Republic revealed moon hoax information today now the Cold War is all over? You decide.

Chapter 42   The Murders

'Three astronauts died in 1967 on the launch pad while testing an early Apollo spacecraft. They were Ed White, Roger Chaffee, and Gus Grissom. These were murdered by NASA because they wanted to expose the hoax to the public. They knew too much and often criticised NASA.'

This 'point of view' (and that's about it), is sustained without any evidence. These astronauts did criticise NASA but so did they all. There was the target of getting to the moon in less than 32 months, the pressure was enormous. The project had so far cost $15,000,000,000 and the moon still seemed a long way off. Many technical problems lay unsolved.

The Apollo 1 crew; Ed White, Roger Chaffee, and Gus Grissom.

Immediately after the fire, NASA convened the Apollo 204 Accident Review Board to determine the cause of the fire. Although the ignition source was never fully identified, the astronauts' deaths were attributed to a wide range of lethal design and construction flaws in the early Apollo Command Module. The manned phase of the project was delayed for 20 months while these problems were corrected. The full Oxygen only capsule was to ultimately to blame. This was replaced with a Nitrogen / Oxygen mix instead to reduce the likelihood of any further fires.

The graves of Virgil Grissom & Roger Chaffee at Arlington Cemetery, Washington DC. Photo by the author 2006.

All the test pilots risked their lives even before joining NASA for the moon program. They dealt with danger every step and knew the risks. Even today, every astronaut fully understands that they may be injured or killed in an accident. Flying into space is dangerous business due to the extreme speeds, pressures and pushing of aerospace technologies to the limits. If the 'Proof' of such apparent murders was so strong then how come nobody has ever produced a case in a court?

It is at this point we demonstrate why the hoax supporters should cease the false & unproven ramblings that could never stand up in a court of law. Families and close friends to these astronauts are still alive today. They constantly hear these accusations of murder rather than how they devoted their lives to pushing forward the limits of knowledge in this exciting field of space exploration.

105

These are the three back-up suits for the Apollo 1 crew on display at the Marshall Spaceflight Center, Huntsville, Alabama. Photo by the author in 2013.

Other astronauts have died in training such as Elliot See and Charles Bassett. Photo by the author 2011.

The false accusations are upsetting surviving friends and relatives. This moon hoax business should stop!

Chapter 43   The Von Braun – A NAZI?

Many people claim that the famous rocket scientist, Wernher Von Braun was a Nazi supporter. They say that this is clear evidence of the moon landing hoax. How this logic works, is a total mystery. But was he really a Nazi supporter?

Perhaps a movie should be made of Von Braun's life...
1930; Von Braun joined a group of amateur rocketeers in Berlin headed by Hermann Oberth. They developed rockets around 4ft to 15ft high and flew to altitudes of a few hundred feet. By 1932, they worked for the army under a research & development contract for rockets in the hope they could find a use for a satellite. Von Braun wrote a letter to Hitler in 1933 expressing the possibility of launching the world's first satellite. It wasn't taken seriously by the High Command structure. A new rocket called the A-2 was built and reached a height of around 7000ft using solid fuel. He knew the way forward was with liquid fuel. At about the same time, achievements were being made in New Mexico, USA, and confirmed that liquid fuel was the only option for Spaceflight.

Wernher Von Braun surrendering to US soldiers in 1945

1936; the Olympic Games in Germany, gave Von Braun a way of approaching Hitler again regarding a satellite. He knew Hitler loved publicity and prestige and appreciated that only the military had the money for such projects. So he wrote to Hitler and this time was listened to. The agreement was to join the Nazi Party, construct a rocket based weapon, then he may have his satellite. Von Braun reluctantly agreed, but put off joining the party as long as he could in the hope they would overlook it. But on 15th May 1937, he was ordered to pay 2 marks and filled out the form. By joining any political party, does that mean the person supports its political views? Anybody can join a political party; even just to obtain cheaper drinks at the social club (as the author of this book has).

1942; the A-3 was flown to a height of 50 miles, Von Braun said to his team "We have just invented the spaceship!" The A-4 rocket was developed and reached 67 miles, the edge of space. Dr Goebbels renamed it the V2; the Vengeance Weapon. Von Braun wanted to add on an extra stage to get it into orbit; Hitler said no and said the target was going to be London.

1944; the war was going badly for Hitler and knew an invasion from England was immanent. Von Braun knew the war was completely lost but convinced Hitler to continue funding the rockets as it was a severe drain on Nazi resources. He was also denounced by a dentist who reported him for making remarks that sounded defeatist about the war effort and von Braun spent two weeks in a Gestapo prison. He was released when military leaders convinced Hitler that he was crucial; the Nazi dictator responded, "I will guarantee you that he will be exempt from persecution as long as he is indispensable for you, in spite of the difficult general consequences this will have."

Hitler finally ordered Von Braun to stop research on rockets and help build the V3; a Super gun. This would fire shells across the English Channel and destroy the harbours and even hit London. He refused; it had nothing connected with space travel and ran into the countryside. He lived by stealing from farms and sheltering in barns etc. The gun was built at La

Coupole without him; this was easily destroyed by a special bomb built by Britain's best war scientist - Barnes Wallis. The Normandy Landings took place; Von Braun headed for the American lines and surrendered after breaking his arm in the process. He later convinced 120 other rocket scientists to surrender to the Americans under Operation Paperclip.

Von Braun hardly ever wore a military uniform. He never trained as a soldier, just wanted to launch rockets into space and perhaps one day land himself on the moon. Hardly any pictures exist of him in a Nazi uniform; just suits as in the image shown or a technician's boiler suit.

Left of the picture; a V2 rocket that led to the development of the Saturn V; Smithsonian, Washing DC. Photo by the author.

Chapter 44   Thirteen Months to the Moon?

A European spacecraft called SMART 1 was launched in September 2003 and eventually reached the moon in November 2004. My goodness how can it take all that time to reach the moon with 21st century technology yet NASA claim to have sent astronauts there in a much bigger and heavier spacecraft in the 1960's? Got you again NASA!

The SMART 1 spacecraft used a revolutionary engine called an Ion Drive. Atoms of xenon gas are electrically charged and ejected from the rear of the craft from a 30mm wide nozzle. This produces a gentle forward thrust of 0.02 kg, roughly equivalent to the weight of a postcard. Just 82 kg of fuel required to get from earth orbit to lunar orbit.

The Apollo third stage that fired the Command Module & Lunar Module to the moon had a thrust of around 115,000 kg and fired for just 11 minutes. This sheer brute force propelled the spacecraft to the moon in just under 4 days. The Smart 1 Ion drive had a thrust of just a few grams. Little wonder it took 13 months to get to the moon. The purpose was to test the concept of travelling around the solar system with very little fuel. As no people were on board, what was the hurry? The project worked perfectly.

The Smart 1 space probe used a very low thrust Ion Drive. It took thirteen months to reach the moon on very little fuel. The Apollo vehicles reached the moon by sheer brute force of the J2 engine on the third stage of the Saturn V. Compare this engine to the Ion Drive engine that is just 30mm wide. They wonder why Apollo can get to the moon in just four days compared to thirteen months? They need to examine propulsion systems to gain the right answer rather than just comparing times to get to the moon.

Photo by the author; Johnson Space Center, Houston TX.

Chapter 45   Why did the Soviets give up?

The Soviet Union made huge apparent strides in the early days of the space program. Sputnik 1 launched first, then Sputnik 2 with a dog, then the first Man in Space with Yuri Gagarin, the first 2 man crew, the first three man crew and the first space walk. But if you analyse each achievement, one stride to the next was actually rather small as the same rocket was used each time. The top section was changed for differing uses and the headline changed with it. But the achievements were making history regardless. Even today in the 21st century, the same R7 rocket booster is still used albeit with upgrade modifications.

The impressive headlines being made panicked the public in the USA. President John F. Kennedy had to qualm the fears of possible Soviet nuclear weapons raining down from space and asked the newly formed NASA what could be done in a few years time to catch up and excel the Soviet achievements.

A manned landing on the moon was the perfect answer. A giant rocket would be needed to launch the crew, a separate Lander and carry enough fuel for a return journey. It was much more likely that the US would develop this ahead of the Soviets. Research and design began immediately on the mighty Saturn V rocket. The Soviets reacted by designing their moon rocket called the N1. They planned to build twelve N1 rockets, but after four test failures that commenced in February 1969, they knew they had no chance of beating the US to the moon. The first Saturn V flew on 9 November 1967 as the Apollo 4 mission and flew perfectly. This was the point where the US edged ahead of the Soviet Union in space.

Clips of the first Saturn V and the first N1 launches are found on the website... www.moonlandinghoax.org

After Apollo 11 landed in July 1969, the Soviets announced a message of congratulations and then later claimed that there was no race to the moon. At the same time they secretly mothballed the hardware and paperwork that was connected to their moon landing attempt.

After the collapse of the Soviet Union, historic records and hardware were gradually released regarding the N1 and the secret moon landing program. The following image shows what could have become the Lander itself. Alexey Leonov would almost certainly have been the No.1 Cosmonaut in line for the lone mission. He made the very first walk in space outside a capsule in March 1965 on the Voskhod 2 mission.

The actual proposed Soviet Lunar Lander on display in the Moscow Aviation Institute. Big enough for one Cosmonaut.

113

The Russians quit the race to the moon almost soon as Neil walked down those nine steps and placed the first human footprint in the lunar dust. The Russian leaders decided that risking a dangerous mission that could lead to a Cosmonaut's death just to be second on the moon was too much to bear. The least damaging option was to claim that there was no race after all and the construction of a space station was their original aim.

The world's first space laboratory was hurriedly built and named Salyut. It was rushed into production and launched in April 1971. It was visited by the Soyuz 10 crew but was unable to attach due to technical problems with the docking latches and returned to earth. The launch of Soyuz 11 in June of that year made a successful docking and became the first occupied space station. The achievement was marred however with the landing. The capsule depressurised during re-entry and all three suffocated. There was such a low weight restriction, that none of the astronauts could wear pressurised suits. Independent investigators concluded that the serous technical faults during this era solely occurred due to the rapid alteration from a moon landing goal to a space station.

The Salyut One space station 1971

Chapter 46 Soviet Tracking

As the Soviet Union were planning there own lunar landings as well as other space missions, they needed to track their spacecraft 24 hrs a day. NASA had asked various radio telescopes around the globe to assist them. As the world rotates the NASA communications based in the USA will often be facing away from Apollo so they requested help from Jodrell Bank in England, Parkes Observatory in Australia etc. The Soviets wanted full control of their tracking so they built ships to have the relevant communications systems that can freely sail around the globe in international waters and not rely on voluntary assistance.

Above is the ship Kosmonaut Vladimir Komarov; named after a Cosmonaut that was killed during his mission. It was a cargo ship but was converted into a Spaceflight communications vessel in 1967. This is one of the ships that not only tracked Soviet space flights, but the NASA missions too. NASA had mastered the transmission of data on a very narrow band width, the Soviets wanted to study how this was carried out in more detail for their own telemetry. It was decommissioned in 1989.

As Neil walked down the ladder on 20th July 1969, the live image was beamed into the Kremlin direct from the ship shown previous. It wasn't shown live to the public but it was decided to eventually to transmit the recovery of Apollo 11 on the 24th July. Viewers had live coverage for the last part of the mission via Moscow TV that hooked into Eastern Europe's Intervision network. They showed the parachutes opening and later the three astronauts on the carrier Hornet. They devoted first two-thirds of the final newscast to Apollo 11 crew and announced that Soviet President Nikolay V. Podgorny had sent a telegram to President Nixon offering "our congratulations and best wishes to the space pilots."

Soviet President Podogorny in 1969; He was President of the Soviet Union from 1965 to 1977; Wikimedia image.

The Soviet Union never doubted the claims of the NASA moon landings. Not only did they have spies in the USA but the live feed from all the Apollo missions from their own advanced receiving stations proved to them beyond any doubt that each mission was genuine. If they did discover something was amiss, they would have shouted 'Foul' as loud as they possibly could for the world to hear. Nothing was ever even whispered.

Chapter 47   The Landing Sites Today

The six landing sites left hardware, tyre tracks and footprints that should be still there today. Through a good quality telescope, in theory, we should be able to see it all. But no telescope on Earth ever has. Not even the Hubble Space Telescope has photographed any landing site features even though it can take photos of faint galaxies billions of times further away. Why is that eh? They weren't there were they?

We still have to remember the Moon is 240,000 miles (400,000 km) away. The largest objects left on the moon are the lower stages of the Lunar Module. These were just 5mt wide, tyre tracks are around 12" (30cm) wide, and the footprints are narrower. Resolution of the telescopes is the key here. From the moon you could try to spot the Great Wall of China. It may be 1000 miles long but the width is more important. From 240,000 miles, you cannot spot anything that is just a few yards across. The 'fact' that the Great Wall of China can be seen from the moon is actually a myth. But high resolution satellite photography from just 300 miles up can reveal it.

Experiment; Get two people to hold up a long piece of thin string. This can be as long as you like, say 20 metres. Each person holds it taught at either end. An observer should walk away say 100 metres. Can the observer see the string? It's long enough but not wide enough.

Now here comes the best bit. A satellite called the Lunar Reconnaissance Orbiter (LRO) has indeed imaged all six Apollo landing sites from low lunar orbit. Apollo 11 is shown first. The crater that Armstrong manually flew over is to the right of the LM. It has been named as Little West Crater.

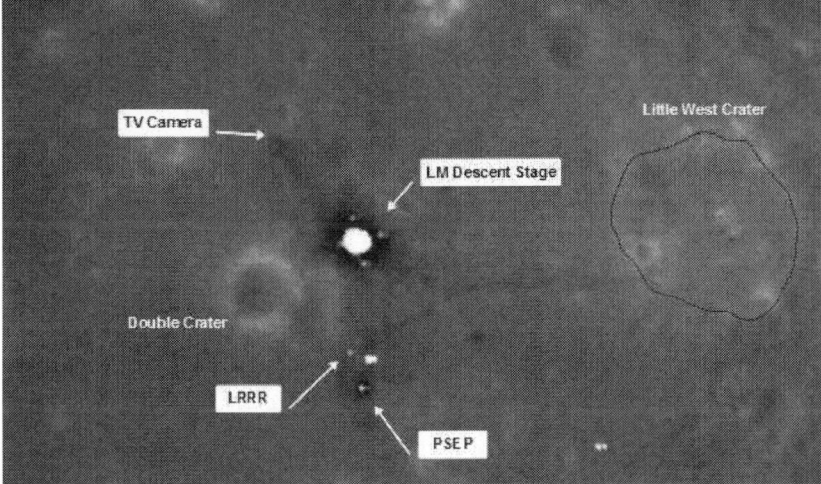

Apollo 11 Site

TV Camera

Little West Crater

LM Descent Stage

Double Crater

LRRR

PSEP

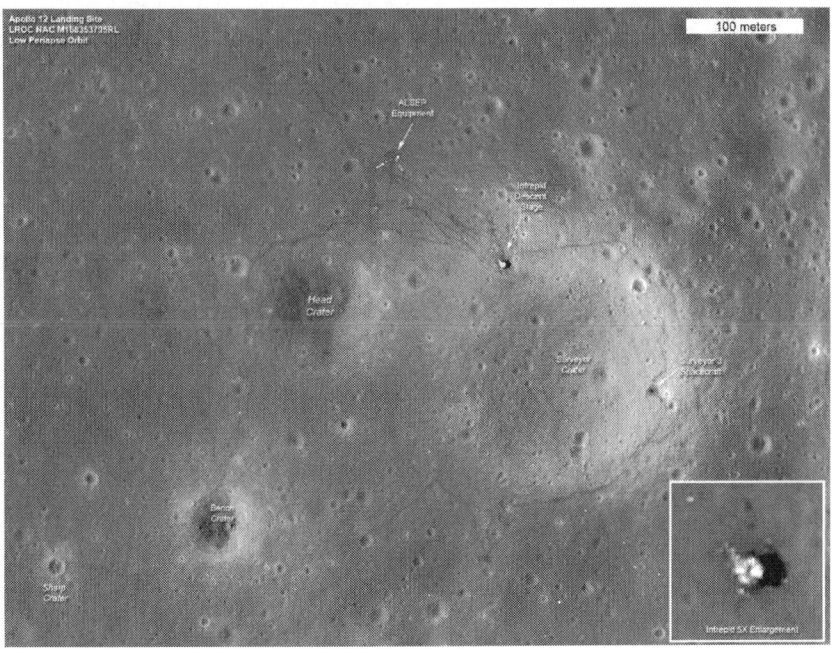

100 meters

Apollo 12 Landing Site
LROC NAC M1B...
Low Periapse Orbit

ALSEP
Equipment

Intrepid
Descent
Stage

Head
Crater

Surveyor
Crater

Surveyor 3
Spacecraft

Bench
Crater

Sharp
Crater

Intrepid 5X Enlargement

The Apollo 12 Landing Site

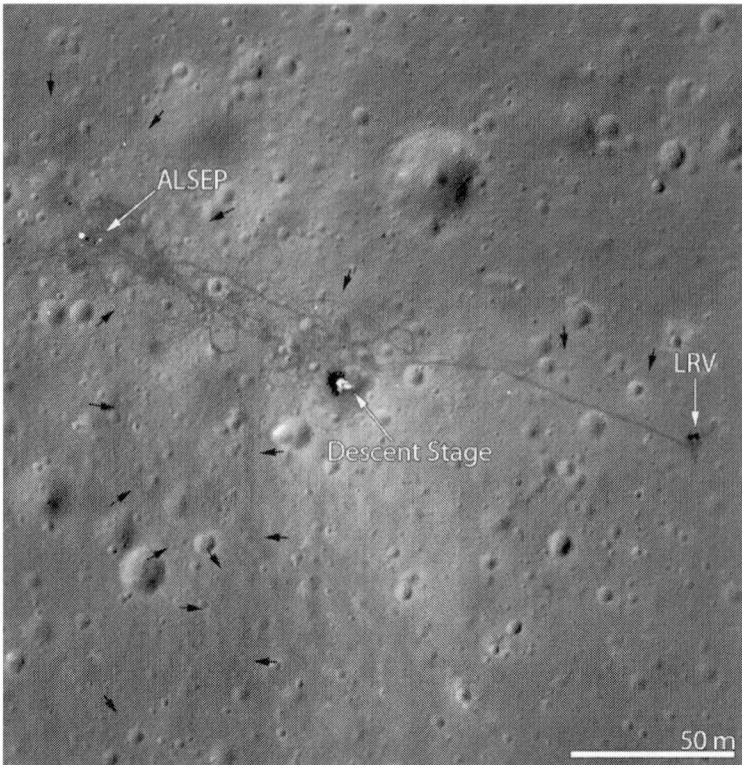

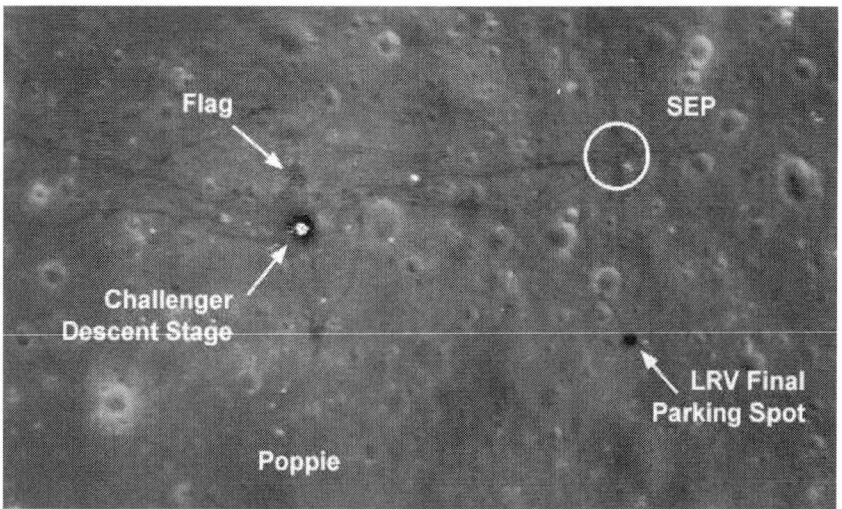

The Apollo 15 & 17 landing sites. I'm sure you get the idea by now. Similar images have been obtained by lunar satellites built by other nations although not quite as high a quality as these.

Chapter 48   Wonky Science

You-Tube is a fantastic place to uncover all kinds of wacky & bazaar beliefs. Ideas from "The Earth is flat", "Our planet is going to get hit by another called Nibiru in 2012", "The earth has tipped over" as well as the Moon Landing Hoax. I never fail to be surprised as to what Wonky Science people believe in. If they just ask themselves a few very simple questions, they would realise their mistake and delete the clip they have published.

Nibiru was supposed to be here and destroy the Earth in 1977, 1986, 1999, 2000, 2001, 2012, 2014, 2015 and on it goes. Sometimes several dates are given in each of those years. Ask the people who post these clips 'Will you give away all your possessions to me as the date approaches?' They never reply.

If the earth was flat then how come the Earth's shadow on the moon during a Lunar Eclipse is round? If we witnessed such as eclipse at moon rise or set, then the earth's shadow would be a straight line. Guess what shape we see? A round shadow!

I have included a few other examples on the website and they will change in time as some realise the error of their ways and delete it. Never fear, there will be plenty more. Don't forget these are all adults and really believe this stuff. If you are at school today and laugh at these examples, then you will do well in science, or at least do better than these people; so give yourself a pat on the back.

Some publishers of such material don't believe a word of their own crazy 'theories.' If the clip has thousands of hits and subscribers, adverts are then placed on it and they get paid. That is their real agenda. I won't sink low enough to emulate them.

Chapter 49   Handling a Hoaxer

Many people are unsure what to believe about the moon landings after watching the Fox TV documentary on the moon landing hoax. All they need to do is take a look at our website and / or read this book. They should then approach a professional photographer, a radiation expert who may work in a hospital for instance, visit an observatory that regularly fires lasers at the reflectors left on the moon, or better still go along to a local astronomy club.

The avid hoax supporters won't bother doing anything. They love to ramble on about one point, then when proved wrong, they start another without admitting they were wrong on the previous point first. After being shown they are wrong on a few items, they then start swearing and running down the US government, praising Lee Harvey Oswald for his assassination of JFK, criticising the Iraq war, claim the 9/11 attacks were planned by the CIA and finally bring God into it. They will then mention the first point all over again and you realise they are not interested in learning anything or simply rather not deal with the real world. Such people are stuck in their little fantasy and are happy to try and depress you. If they succeed, they are overjoyed. Just keep your cool. They will never turn to reality, so it's best to walk away, or change the subject.

Bart Sibrel, a well known hoax believer kept pestering Buzz Aldrin about the moon hoax. Bart turned up at his home, his office, when he was dinning with friends, pestering him, trying to get Buzz to admit to the hoax. He lost his patience on one final occasion and threw Bart a right hook on his chin. Wack! Bart took Buzz Aldrin to court for assault but lost the case, Bart had to pay all the fees.

I don't suggest you follow this course of action.

I have begun to paste some replies from conspiracy supporters on the website. I have omitted their names to avoid their full embarrassment but I have included a couple here...

'Hey Peter, soon almost all religions with their origins in ancient Babylon will be removed from Earth. All off them. Take a look at the news. See how nations are gathering together to attack religions such as Islam. Japan are furious that their people are being slaughtered. France likewise. Others are involved. It's a war on religion. Terrorism is an excuse. You will soon lose your freedom. NASA was founded by lying nazi occultists. It's run by lying Masonic occultists and deceptive Jewish kabbalists. They faked the lunar landings to gain world dominance. It was political. These people are members of a Babylonish religious cult. This cult MUST be removed along with other false religion. If you are involved, get out now. Otherwise, remember my words...'

Come on man, "ask a professional and take their word for it" who are you kidding? Are you a freemason? Why lie to people? Do you believe the bible good sir? I do. That's why I've come to my conclusion that the earth is actually flat and we've been lied to in order to cover it up. I know it might sound crazy, especially if you aren't a freemason but anyways I figured I'd give you a dose of "crazy" today to make you smile. The truth is not hidden it's been in the bible this whole time.

My standard kind reply is 'I think you need professional help!'

Many hoax supporters will read this book, examine the associated website and never admit they have been wrong about anything. They will leave bad feedback and they will pull out another picture they don't understand and ask questions about that instead. As I hadn't included it, they will claim I have something to hide or NASA had kept it secret and it has 'slipped out.' They will expand on an obscure point for months. Once someone has responded from one source or another with the correct answer to their problem, they know it sounds right but they will start all over with yet another 'mystery.'

123

Chapter 50   Why they do it!

Why do people manufacture conspiracy theories? Some do turn out to be true. Big businesses create secret deals (such as the recent Volkswagen emission test scandal) or cover mistakes, wars generate secret operations; a small number of people can create a massive lie for millions to be duped. Perfect examples are demonstrated in World War Two movies... 'The Man That Never Was', 'Monty's Double' and 'The Eagle has Landed' although the last one was greatly exaggerated, nevertheless all based on true operations.

But in science, theories can be easily tested, then thrown out such as the Moon Landing Conspiracy or 'the Moon is one gigantic Alien Spaceship.' (I may deal with that one next). Hopefully by studying these pages everybody by now has accepted the Moon Landings were for real or at least for those that already knew that, it gives you more ammunition to use on hoaxers. But this hoax business still just doesn't vanish quietly... Why?

After dealing with this for over 15yrs, I have found several key answers...

There are many people who would love to be a scientist but just don't want to put in the time. They feel if they run down some scientists with their work, they sense they are above them without bothering to study.

Some people crave attention. Regardless of who they upset, they either create a conspiracy from very tenuous 'facts' or go along with and promote an existing one. With the easy options on the internet to get their name around, they get the attention they seek; although they are promoting nonsense. There will always be some that will believe them, mostly people in a similar position to their own.

124

There are those who hate their government. They love promoting anything that embarrasses it even though the method of promoting false conspiracies end up just embarrassing themselves without realising it.

Some people dislike themselves and their own personal situation. These will promote anything that run down top people - scientists, politicians, business leaders etc. Again, they feel that they are above them without working for it.

The most common; people who know little about science, get taken in by the conspiracy and just get led by it. They may not know an expert in the field in order to gain the correct answers. This book is perfect for those people. It should contain all that is required to realise the real truth about Project Apollo as well as gain skills to investigate other conspiracies and not follow the nonsense.

The rest will probably continue to drone on, gaining attention, a little fame etc. Most know they are wrong, but will never stop regardless of how stupid they sound. If they shout loud enough, some will always believe them; this in turn fuels their strange occupation. Many claim to be science experts, but a few carefully worded questions to them will show their true colours. Tearing apart these conspiracy theories do train your mind to be more analytical. It is important to do this and make the truth public as such theories upset many people; in this case astronauts, the NASA employees and families. Many do get harassed in their private lives from the public.

The September 11 2001 attacks also have an associated conspiracy; that it was an inside job by the CIA. This accusation is far worse than the moon hoax. Thousands of people died that day, the USA, Britain and others went to war in Afghanistan to try to prevent another attack. But these conspiracy people undermine the memory of all those killed by this ridiculous CIA plot that also didn't happen.

125

The basis of the conspiracy surrounds the destruction of Building 7. After the Twin Towers collapsed, Building 7 just fell down and was recorded on video from the back. It didn't seem to be on fire and no planes hit it. So the conspiracy people invented the idea that the CIA evidence was there so the building had to be destroyed. Firstly why demolish a multi-story building when a £30 shredder would have done the same job? Secondly, the building collapse would not guarantee the destruction of this fictitious evidence. But most important of all, the building was indeed on fire at the front on several floors. Hundreds of witnesses saw it including police officers and fire fighters.

After the collapse of the Twin Towers, Building 7 was evacuated and was left to burn. The structure weakened and collapsed. The only cameras that were still operating were those behind the building that couldn't see the raging fire. No mystery to solve in the first place.

I am sure many avid conspiracy supporters will read this book and not be convinced of where there they are going wrong. They will never admit they are in error about anything. They will ignore every single point answered and just move on to some other mystery they don't understand.

Others will insist that there are two sides to every conspiracy. I disagree. There can only be one; the correct side.

Perhaps the spread of such misguided gossip should be made illegal as it is in Germany promoting WW2 Holocaust Conspiracies.

Chapter 51   Other Daft Conspiracies

False conspiracies upset huge numbers of people in a very personal way. The Internet today makes it too easy to invent one, publish it, and get millions to follow ridiculous lines of thought. People who are not experts in that particular field, know no better and spread the idle upsetting gossip further until it behaves like a virus. Here are some more examples and a starter of how to pull it apart. I won't do it all as it would take a whole book like this one to do it fully for each conspiracy. I just don't have the time but would love to. However the principal of simple realistic investigation is the same for all of them...

The Oklahoma City Bombing... The explosion was an even force and yet the damage to the building was uneven. Also witnesses inside said they had the damage occur first, then heard the explosion outside after. So there must have been a bomb placed inside as well as outside! The CIA probably did it, they have explosive experts!

The bomb blast was an even force but the building strength was not. Open office space is weak with thin internal walls, but stairwells & lift shafts are extra strong. So those parts survived more readily than open office floors. The shock wave from any explosion is far faster than sound, so damage is done first then hear it after just as with fireworks or lightning. Tim McVeigh became an explosives expert via his US Army training and his survivalist friend, Terry Nichols. NO CONSPIRACY!

9/11 Attack on the Twin Towers... The second tower hit came down first, and windows shattered ahead of the collapsing floors. They didn't collapse in the way they should have done after being hit by planes. The steel should have slowed down the collapse as some other buildings have. Building 7 collapsed later without it being on fire as seen from CCTV. The CIA did it again as an excuse for war.

The second plane hit 16 floors further down the building so a much larger force was bearing down on the damaged part. If the planes hit higher up, they may not have collapsed at all. As each tower collapsed, air trapped inside rushed down stairwells and lift shafts blowing out the windows. It wasn't explosives being detonated. How many 110 story towers have we witnessed collapsing from airline hits? What way were they supposed to come down? Downwards is one way only! The sheer mass and energy of all this debris couldn't be stopped by the iron works that held the buildings up, these are the tallest buildings ever to have collapsed so this event cannot be compared to other smaller demolitions. Building 7 was on fire on the other side of the said CCTV footage. Hundreds of witnesses and other cameras recorded it. It was abandoned by the fire-fighters. It didn't just fall down. NO CONSPIRACY!

Hiroshima 1945... Japan was going to surrender anyway. The Atom Bomb was only used on Hiroshima & Nagasaki as an experiment to see how such weapons work on real Cities, and for the USA to get their 'monies worth' from all the research.

Japan had every opportunity to surrender. President Truman told the Japanese military in advance that such a weapon existed and would use it if no unconditional surrender was made. No response from Japan for several days, Hiroshima was bombed. They were given further chance to surrender but still nothing. All diplomatic channels remained open to the USA as well as other allies such as Great Britain; they were still fighting the Japanese in Burma and Malaya. A radio announcement would have sufficed. Nagasaki was bombed two days later, still no surrender. Russia declared war on Japan and began moving troops & ships. Only then did Japan surrender unconditionally to the USA or face the possibility of being invaded by Russia while the USA produced more atomic weapons. NO CONSPIRACY!

Satellites are fake! Navigation from space is secretly done by Mobile Phone masts and communications are made by cables under the sea. All pictures from space are faked!

These people should get out more... literally. All they need to do is pop out under a clear dark sky and look for moving stars that majestically cross the night sky. Guess what these are? Yes you got it... Satellites! So how do ships, boats and aircraft use navigation via mobile phone masts in the middle of the Pacific Ocean? I'm not even going to bother with the rest. They should go to a rocket launch site and see a satellite commence its mission. I have an entire book on the subject; details are in the last chapter. NO CONSPIRACY!

Nibiru is about to hit Earth.
There is a fictitious planetary system in the solar system that regularly comes near the earth every 36,000 years. All you need to do is ask them for the RA & DEC co-ordinates so we can have a look at it through telescopes. They never respond fully but just say the whole system is invisible and can only be seen in the infra-red. So why doesn't the Sun illuminate the objects as it does with Jupiter? It's also an infra-red object. They can

never answer that one. They sometimes reply that it can only be seen from the South Pole. But any astronomical object seen at the south pole can be seen all the way up to the equator just as the Southern Cross constellation can be viewed; another crackpot conspiracy.

The Earth is Flat ! They say that Gravity is a hoax too. It's really the Earth flying upward at an increasing speed of 9.8mt/s giving us a false sense of gravity. Well that means we will be travelling faster than light within a year. Hold on to your hats, we are now travelling at millions of miles per second.

How do pilots & sailors navigate around the globe if all the maps are wrong? They say the UN flag shows the flat Earth map. How are you supposed to create a round object on a flat flag? It is not a geographical map anyway but a political one instead. Antarctica isn't on it as it has no government there. Each continent has been altered in area for equal representation. It is not meant to be a map.

A Lunar Eclipse will show the Earth's shadow. If the Earth was flat then an eclipse seen at moon rise or set will show a straight line shadow. It never does. The shadow is round regardless of the height of the moon in the sky. It can only work if the Earth was a sphere; the Greeks discovered this over 2000 years ago.

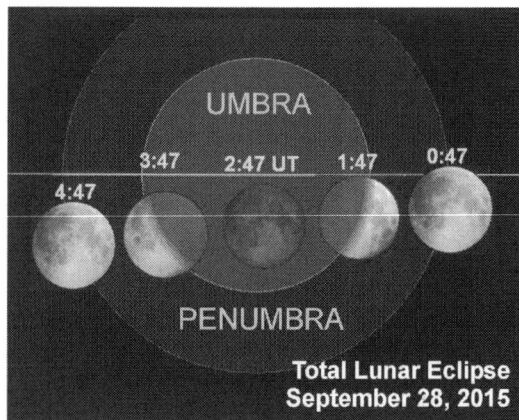

Credit to Sky & Telescope Magazine. The Earth's shadow is ALWAYS round regardless of the moon's position.

And how can we observe stars rotating around the North Celestial Pole (North Star) and rotate around the South Celestial Pole too if we are living on a flat rotating disk with one centre of rotation? This is probably the nuttiest idea I have ever heard so far. I just can't understand how people can really believe this flat earth nonsense. It defies all logic.

The Moon is a hollow Alien Spaceship, Atomic Weapons don't exist, the Earth has tipped over, Hitler survived the bunker or The Titanic didn't really sink!

I am not even going to waste any more of your valuable time here… these conspiracy people don't know when to stop or who to upset next.

Returning to the subject of Apollo, there are those who believe the opposite of this book and that all the Apollo missions were real including the Sci-Fi movie Apollo 18. They insist we didn't go back after Apollo 18 as there is intelligent life there that we upset. For the record; Apollo 17 was the last true moon landing in December 1972. Folks; Apollo 18 was just a movie. Ask the director, film crew & actors etc. There were hundreds of personnel involved with the making of it. There are no aliens on the moon.

131

Chapter 52   Some True Conspiracies

Big business can induce massive cover-ups such as the recent Volkswagen Emission Test, warfare can induce the same. But conspiracies that involve science projects can easily be resolved.

Movies have been made about some true conspiracies that occurred during World War Two;

The Eagle Has Landed... around 10% true (Film was greatly exaggerated to make a more exciting story)

Where Eagles Dare... around 10% true (Film was exaggerated to make a more exciting story)

Monty's Double... around 60% true (Based on a reliable internal information leak before full admission)

Valkyrie... around 90% true (More accurate as the film was made long after the truth was revealed). One of the very few errors was when Hitler's burning body was saluted from outside the bunker; it was saluted from inside the bunker with the door closed as the heat from the fire was too intense.

The Man that Never Was... around 80% true (Based on a reliable internal information leak before full admission).

A cold war movie;

A family of Spies... around 95% true (More accurate as the film was made long after the truth was revealed).

Capricorn One... 0% true ! (This is the most accurate one to put a figure on)

Chapter 53   Visiting Apollo

All the Apollo Command Module spacecraft built are on display in various museums (Apart from Apollo 1 due to the fire; in respect of the deceased). We have listed the current sites where the main Command Modules can be viewed.

Apollo 6	Fernbank Science Center, Atlanta, GA
Apollo 7	Frontiers of Flight Museum, Dallas, TX
Apollo 8	Museum of Science & Industry, Chicago, IL
Apollo 9	San Diego Aerospace Museum, CA
Apollo 10	Science Museum, London UK
Apollo 11	Smithsonian, Washington DC
Apollo 12	Air & Space Center, Hampton, VA
Apollo 13	Kansas Space Center, Hutchinson, KS
Apollo 14	Kennedy Space Center, FL
Apollo 15	US Air Force Museum, Dayton, OH
Apollo 16	Marshall SF Center, Huntsville AL
Apollo 17	Johnson Space Center, Houston TX
Apollo Skylab 2	Aviation Museum, Pensacola, FL
Apollo Skylab 3	Science Center, Cleveland, OH
Apollo Skylab 4	Smithsonian, Washington DC
Apollo / Soyuz	Science Center, Los Angeles, CA

Author standing by Apollo 12. Further details with more posing pictures and direct museum links are on the website.

Chapter 54   The Author's mission

Peter Bassett FRAS became interested in Spaceflight in 1968 with the Apollo 8 mission. He saw the first live broadcast from Lunar Orbit on 24 Dec. Mum pulled back the curtains and they saw the crescent moon through the window at the same time as watching the astronauts on TV.

He has been interested in photography since 1977 when he purchased his first camera with paper-round money. It was fully manual and learned various aspects of how a camera works regarding exposure, depth-of-field etc. He took pictures of stars, the moon, planets, (Astro-photography) so he knows how to correctly produce such images. He turned his photographic skills into a business by Aug 1981. During this period he gained an interest in 8mm film making and completed two amateur sci-fi movies full of special effects. So he understands lighting, multiple exposure techniques, what was capable and what wasn't with film at that time. He studied the effects on movies such as 2001 A Space Odyssey, Silent Running, Star Wars, Close Encounters etc. A film company was then created in 1987 - 'The Movie Factory.' The company didn't flourish as hoped but he did gain professional experience within the field of special effects.

In 1994, Peter became a professional lecturer in Astronomy. He uses a mobile planetarium to lecture at schools and colleges on Astronomical & Spaceflight related topics. By 1997, he began to hear from pupils about the moon hoax conspiracy. "Flags cannot fly on the Moon" and "Why are there no stars in the pictures?" were shouted out often. On rare occasions, these comments were from teachers. That's when he decided something needs to be done about this. Peter wrote a book proposal, but couldn't find a publisher to take it on, some hinted they would if he supported the hoax instead; that was very concerning.

With each trip to the USA, Peter takes images of space centres, museums etc to help illustrate the points mentioned in this book and website. During the process he has met most of the Apollo astronauts too.

The author posing by the Quarantine Facility that housed the Apollo 11 crew after splashdown on board the USS Hornet. This is docked as a museum in Alameda, San Francisco.

The author, his wife Amanda and Buzz Aldrin in 2006.

Producing this work didn't require any backing; it just cost a little time and £2.99 a year to hold the web address. If he has steered even one person in the right direction toward real science and how it is done, he is happy. He dislikes those that are fundamentally lazy; they guess at science, assume the guesses are correct then publish them. People who know no better follow it and publish their own nonsense on the internet.

This trend needs to be halted before the education process about history & science is watered down to a dribble of nonsense. How are we going to solve the world's problems otherwise?

There are two main kinds of people in this world; 1) those who think the glass is half full. 2) Those who think it's half empty. Peter is in a third category; the beer is off and he goes back to the bar to get another for free.

We all need to get real instead of this crazy world of imaginary & potty conspiracies.

The world faces huge problems today; Famine, Energy & Material Shortages, Climate Change, Destruction of the Natural World, New Diseases etc; all can be solved by new realistic science and policy. By increasing the number of people involved, we increase our chances of survival or fall upon very hard times and remain stuck here on this tiny Planet Earth for centuries.

The 21st century is a unique era where we get this choice.

Chapter 55   Other Books by the Author

For the very latest listing of all signed books and versions, use www.astronomyroadshow.com They all have a supporting website.

This book is available as a paperback in colour or B&W, an E-Book via Amazon. Internet links immediately accessible.

Satellite Spotting and Operations Handbook for Beginners. A shorter version called 'Satellite Spotting for Beginners' is available too.

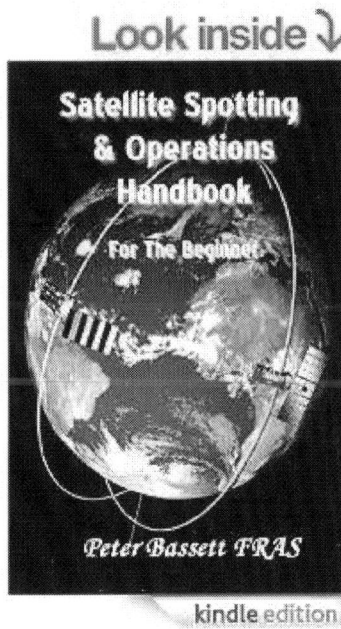

Over 40 chapters, 250 + pages, dedicated website

www.satellitespotting.co.uk

This book is available as an E-Book via Amazon. All the internet links will then be immediately accessible.

Look inside ↓

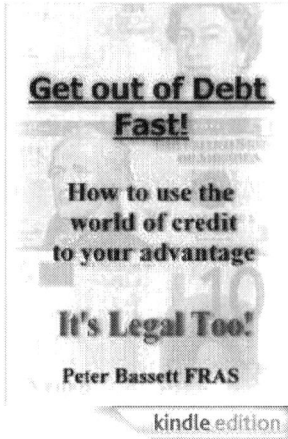

Get out of Debt Fast!
Available as an e-book only at present.

www.savetheplanet.org.uk

Also out in print and e-book... www.hedgehoghome.co.uk

Look inside ↓

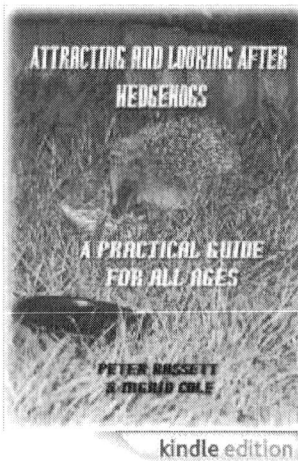

Attracting and Looking after Hedgehogs
A book written for pupils from age 10 to adults. Available as an e-book or printed book via

www.hedgehoghome.co.uk

13980252R00078

Printed in Great Britain
by Amazon.co.uk, Ltd.,
Marston Gate.